ACTA MECHANICA
SUPPLEMENTUM

1

O. E. BARNDORFF-NIELSEN AND B. B. WILLETTS (EDS.)

Aeolian Grain Transport 1
Mechanics

C. Christiansen, J. C. R. Hunt, K. Hutter, J. T. Møller,
K. R. Rasmussen, M. Sørensen, F. Ziegler
(Associate Editors)

SPRINGER-VERLAG   WIEN   NEW YORK

Prof. Ole E. Barndorff-Nielsen

Department of Theoretical Statistics, Institute of Mathematics, University of Aarhus,
Aarhus, Denmark

Prof. Brian B. Willetts

Department of Engineering, University of Aberdeen, Kings College, Aberdeen,
United Kingdom

Printed in Germany by Maxim Gorki - Druck GmbH, Altenburg

Printed on acid-free paper

With 79 Figures

ISSN 0939-7906
ISBN 3-211-82269-0   Springer-Verlag Wien — New York
ISBN 0-387-82269-0   Springer-Verlag New York — Wien

Dedicated to the memory of Ralph Alger Bagnold

Foreword

Wind erosion has such a pervasive influence on environmental and agricultural matters that academic interest in it has been continuous for several decades. However, there has been a tendency for the resulting publications to be scattered widely in the scientific literature and consequently to provide a less coherent resource than might otherwise be hoped for. In particular, cross-reference between the literature on desert and coastal morphology, on the deterioration of wind affected soils, and on the process mechanics of the grain/airflow system has been disappointing.

A successful workshop on "The Physics of Blown Sand", held in Aarhus in 1985, took a decisive step in collecting a research community with interests spanning geomorphology and grain/wind process mechanics. The identification of that Community was reinforced by the Binghampton Symposium on Aeolian Geomorphology in 1986 and has been fruitful in the development of a number of international collaborations. The objectives of the present workshop, which was supported by a grant from the NATO Scientific Affairs Division, were to take stock of the progress in the five years to 1990 and to extend the scope of the community to include soil deterioration (and dust release) and those beach processes which link with aeolian activity on the coast.

The meeting satisfied these objectives and proved most stimulating. Presentations described both completed studies and work in progress, drawn from many countries and several disciplines. They provided stimulating discussions, the fruit of which will become evident during the next few years. The standard of presentations was generally high and most of them are represented in this collection of papers. Authors have had opportunity to incorporate the outcome of discussion where this was appropriate. So much good material was produced that the papers extend to two volumes.

The subject matter has been structured so that papers appear in cognate groupings because it was felt that this would assist readers, particularly those for whom this is a new field. The two volumes are, accordingly, headed respectively "Mechanics" and "The Erosional Environment", and each contains sub-sections, as can be seen in the table of contents. However, we found difficulty in assigning some sub-sections to a particular volume and some papers to a particular sub-section. This we regard as an encouraging sign that the boundaries between different views, associated with different disciplines and fields of application, are becoming more pervious, and cross-fertilisation more active.

The study of aeolian processes is at a truly exciting stage: the primary applications — desertification, coastal defence, dunefield behaviour — have never been more important. We hope that many people will read these papers, find them interesting and challenging, and will correspond with the authors, to the benefit of the development of understanding of the subject.

During 1990 the deaths occurred of both Ralf Alger Bagnold and Paul Robert Owen, two outstanding contributors to this subject, Bagnold pre-eminently so. Both of them attended the 1985 Workshop, and both would have attended this one had their health permitted. The two deaths greatly diminish the scientific community and, as you will see we have respectfully dedicated one of these volumes to the memory of each of them.

O. E. Barndorff-Nielsen
B. B. Willetts

January 1991

Contents

Acta Mechanica (1991) [Suppl] 1: 1—19
© by Springer-Verlag 1991

A review of recent progress in our understanding of aeolian sediment transport

R. S. Anderson, Santa Cruz, U.S.A., **M. Sørensen**, Aarhus, Denmark,

and **B. B. Willetts**, Aberdeen, United Kingdom

Summary. We review recent progress in our understanding of aeolian sediment transport, with emphasis on work published since 1985. The current conceptual model of sediment transport is discussed at length, with attention given to problems of definition that have arisen. We discuss in depth the collision (grain impact) and aerodynamic entrainment (initial motion) processes. The effect of the evolving population of moving grains on the wind (the wind feedback mechanism) is treated in the context of recent modelling of the self-regulating saltation process. The link between saltation and suspension is discussed briefly. We conclude by outlining future research directions that must involve a greater symbiosis of experimentalists and theoreticians, working both at the grain and the bedform scales.

1 Introduction

Important progress has been made in the recent few years in our understanding of aeolian sediment transport. In the present paper we review those results that we find most significant. We focus on the time after the Workshop on the Physics of Blown Sand held in 1985 at Aarhus University. The understanding of aeolian sand transport in 1985 is well reflected in the Proceedings of the Aarhus Workshop [12]. Work published before 1985 will be discussed when this seems natural or necessary.

The considerable progress in the mathematical models of aeolian sand transport has not yet lead to more reliable predictions of transport rates. It is, moreover, questionable whether this approach will ever yield universal transport rate formulae. We discuss this problem further in Section 2. The significance of the mathematical models is that they have lead to a much deeper understanding of the physical processes involved in determining the transport rate. This is important for a correct interpretation of geological processes, might be useful in constructing protection against drifting sand, and gives the insight necessary for developing sensible empirical transport rate formulae under given field conditions.

The paper is organized as follows. In Section 2 we outline the current conceptual model of aeolian sand transport. The collision process taking place when a sand grain impacts the surface is discussed in Section 3 together with the related transport modes' reptation and creep. The initiation of sand movement is considered in Section 4. In Section 5 we discuss the self-regulatory nature of the saltation process and its role in establishing the dependence of transport rate on wind speed. The link between saltation and suspension is discussed and formulated mathematically in Section 6. In the final Section 7 some suggestions for future research are given.

2 Outline of the current conceptual model

Our conceptual understanding of the process of aeolian sand transport has progressed substantially in recent years. R. A. Bagnold's [9] classical perception of the process as a cloud of grains leaping along the sand bed, the grains regaining from the wind the energy lost when rebounding, is the basis for the new developments, but it has been modified and sharpened at important points.

When the wind blowing over a stationary sand bed becomes sufficiently strong, some, particularly exposed, grains are set in motion by the wind. Presumably, some grains are lifted fluid dynamically by the difference between the pressure below and above them, while the rest are pushed forward by the wind before they take off, see the discussion in Section 4. Once lifted free of the bed, the grains are much more easily accelerated by the wind. Therefore, as they return to the bed, some of the grains will have gained enough forward momentum that on impact they rebound and/or eject other grains.

We define *saltation* as the transport mode of a grain capable of rebounding or of splashing up other grains. This is nothing but a slight sharpening of Bagnold's classical notion. We can think of the saltating grains as the high-energy subpopulation of the grains in motion. In the initial period after the wind speed has exceeded the critical value, the sand transport is being built up and the number of saltating grains resulting from one impact is, on average, larger than one. Therefore, the number of saltating grains increases at an exponential rate. We call the average number of saltating grains, rebounding grains as well as ejecta, resulting from one impact the *mean replacement capacity*.

As the transport rate increases, the time-averaged vertical wind profile is modified due to the considerable extraction of momentum from the wind by the grains in motion. The wind speed drops so that the saltating grains are accelerated less and impact at lower speeds. As a result the mean replacement capacity decreases. When the mean replacement capacity has fallen to one, an equilibrium is reached. This equilibrium state is only stationary in a statistical sense. The number of saltating grains fluctuates around the equilibrium value. The mechanism described here is the basic idea behind several mathematical models for aeolian sand transport: Ungar and Haff [52], Anderson and Haff [6], [7], Werner [54], [55], McEwan and Willetts [32] and Sørensen [49]. Computer simulations by Anderson and Haff [6], [7] indicate that the equilibrium is reached very quickly: in $1-2$ seconds.

It is possible that the mean replacement capacity is also lessened because the characteristics of the impact process are modified by the increasing bombardment of the sand bed and by the high concentration of low-energy ejected grains in the air close to the surface. If so, this effect would mainly be seen at high wind speeds. It may also be the case that in the equilibrium state a few grains are still dislodged into saltation by fluid lift. The equilibrium value of the mean replacement capacity would then be slightly smaller than one.

The idea of an equilibrium state of aeolian sand transport is, of course, not new. It also plays an important role in the models of Bagnold [9] and Owen [39], but these authors concentrate attention on the momentum which the grains extract from the wind and transfer to the bed on impact. The horizontal force that the grains exert on a unit area of the bed is called the grain borne shear stress. The difference between the shear stress in the logarithmic layer above the saltation cloud and the grain borne shear stress is called the air borne shear stress. Only under steady, uniform conditions can the air borne shear stress be expected to equal the shear stress of the air flow. Owen hypothesized that an equilibrium is reached when the air borne shear stress at the bed equals the shear stress at the impact threshold as defined by Bagnold [9] (p 32). Bagnold supposed the grains to transfer all momentum from the wind to the bed at equilibrium, see [9] (p 65). Empirical estimates of the grain borne shear stress by Jensen and Sørensen [29] and Sørensen [47] as well as computer simulations by Anderson and Haff [6], [7] and Werner [55] indicate that the equilibrium value of the air borne shear stress at the bed is not as assumed by Bagnold and Owen, see also [49].

The cloud of grains transported in equilibrium with the wind does not consist of saltating grains only. On impact the saltating grains splash up a number of grains most of which will not saltate in our definition, i.e. their impact speed is so low that on impact they can not rebound or eject other grains. Ungar and Haff [52] proposed the name *reptation* for the motion of these low-energy ejecta. This word is derived from the latin verb reptare meaning to move slowly. The notion reptation has in the recent years played a very important role in the theory of aeolian sand transport. An example of this is the new model for aeolian ripples by Anderson [2]. The reptating grains constitute a very considerable fraction of the cloud of sand in transport, but they contribute relatively little to the transport rate because of their short jump lengths. The importance of the ejecta lies in the fact that a small fraction of them jumps high enough to be accelerated by the wind and thus start saltating. After a few favourable impacts such grains become high energy grains. The equilibrium is therefore not maintained because a saltating grain jumps on forever. Rather a saltating grain has a certain small probability of not rebounding which is counterbalanced by the — also small — probability that a dislodged grain receives enough energy to start saltating. The physics of the cloud of blown sand has, incidentally, an analogy with a biological population controlled by and fluctuating around the carrying capacity of its environment, see [5].

The differentiation between whether a grain with a given launch velocity is saltating or reptating is in practice not obvious. That there is a very substantial difference between the distribution of the launch velocity for rebounding and ejected grains has been clearly established by Willetts and Rice [59]. The problem is that a considerable fraction of the ejecta has a non-negligible probability of rebounding or ejecting other grains, and that there is rather little knowledge about how this probability varies with the launch velocity. Therefore, it is at the moment impossible to say exactly for which launch velocities a grain can not rebound or eject other grains on impact. A further problem is that there is more than one definition of reptation in the literature. Werner [55] defined the motion of a grain to be reptation if on impact the grain will on average be replaced by at most a large fraction (e.g. 0.75) of a grain (including the event of rebound as well as that of ejection). With this definition, the population of reptating grains is larger than in the definition given above, but not easier to determine in practice. Sørensen [49] avoided the practical problem of distinguishing between reptation and saltation by identifying the population of reptating grains with the population of all ejecta — a larger population than Werner's.

It is obvious that the impacts at the surface are of crucial importance for the manner in which the equilibrium state develops, for the composition of the cloud of grains in transit and, in particular, for the magnitude of the transport rate. These events are stochastic in nature and can only be described in terms of a probability law conditional on the momentum vector of the impinging grain. Ungar and Haff [52] summarized the outcome of an impact event in the number density of the launch velocity vector and called this density function the *splash function*. The splash function includes the velocities of the rebounding grain as well as of the ejecta and integrates to the mean number of grains leaving the bed as a consequence of an impact event with the given impact velocity. The entire probability law specifying the probability with which the impinging grain rebounds, the distribution of its take-off velocity, the distribution of the number of grains ejected and the distribution of their launch velocities is also very important, see [49]. We propose that the entire probability law be called the *splash law*. Here law means probability law and not a law of physics. Obviously, the splash law is a very complicated object about which there is only limited knowledge at the moment. It is, however, being studied intensively by a variety of methods; see Section 3.

The net transport due to rearrangement of the sand bed caused by saltation impacts also contributes weakly to the transport rate. So do the grains that slide down the lee side of sufficiently steep ripples. The physics of these types of transport differ radically from that of saltation and reptation in that, apart from gravity, forces that are due to enduring contacts between grains dominate their motion rather than fluid forces. We propose that the classical notion *surface creep* be redefined to designate the motion of grains whose displacement is not affected directly by wind forces. The amount of sand transported in this way is undoubtedly significantly smaller than what is transported as surface creep according to Bagnold's [9] rather loose definition, which encompasses reptation as well as surface creep in our sense. Creep traps like the one used by Bagnold [9] and modifications thereof will, by their very construction, mainly catch reptating and saltating grains. Measurement of surface creep in our restricted sense is a delicate matter that has not yet been attempted.

The transport mode of saltation can be subdivided into two types. Most grains follow ballistic trajectories determined by the time-averaged wind profile and are not influenced by the turbulence of the wind. Motion of this type was called *pure saltation* by Jensen and Sørensen [28]. The trajectories of some small grains are modified somewhat by the turbulent fluctuations of the wind velocity. Nalpanis [36] proposed the name *modified saltation* for this type of transport, which is considerably more difficult to model than pure saltation; see Hunt and Nalpanis [26] and Anderson [3]. See also Section 6.

Sometimes smaller grains are lifted far away from the surface by a vertical gust and are transported over long distances without contact with the sand bed. This transport mode is traditionally referred to as *suspension*. Modified saltation differs from suspension in that a saltating grain receives most of its upward momentum from frequent impacts with the sand bed, whereas a suspended grain is carried solely by the turbulent eddies. Presumably there is a continuum of grain behaviour from pure saltation to prolonged suspension.

Mathematical models which quantify the conceptual model of aeolian sand transport discussed above are still far from the point at which reliable predictions of transport rates can be made. Under assumptions of a homogeneous sand bed, a stationary wind, and sand of a single size, predictions can be made ([6], [7], [32], [49], [55]) and will presumably be improved in the relatively near future. However, problems remain in applying the results under field conditions. The wind is gusty, the terrain in-homogeneous, and sand of a single size is virtually non-existent in the field. In particular, it is very difficult to measure correctly the relevant aspects of the wind field moving the sand. The several problems concerning the determination of sand transport rates in the field are discussed in [45] and [42].

There is also the contrary problem that there exist very few sets of reliable simultaneous measurements of the characteristics of the wind and the sand transport rate that can be used to constrain or validate the mathematical models. Numerous wind tunnel experiments have been made, but in recent years it has been increasingly realized that the construction of a good wind tunnel for sand transport studies is very difficult. Secondary flow and low frequency fluctuations are difficult to avoid, and it is even more difficult to obtain even an approximate equilibrium between the measured friction velocity of the boundary layer and the actual shear stress. These and other problems are discussed in Rasmussen and Mikkelsen

[43]. Also the design of efficient sand traps is not a trivial problem. A simple Bagnold sampler works well at a distance of a few centimeters from the sand bed, but close to the bed it grossly underestimates the transport rate. This fact, which was demonstrated by Rasmussen and Mikkelsen [44], may be the reason that the importance of the reptating population has only recently been recognized. Using an advanced iso-kinetic sampler, Rasmussen and Mikkelsen demonstrated that the transport rate close to the bed is considerably larger than is predicted by extrapolating down to the bed the log-linear transport rate profile found by Williams [64] and many other investigators. The extra transport close to the bed is undoubtedly due to the reptating grains, which do not jump far, but are very numerous.

To conclude this section on a positive note, let us point out that there are new possibilities for obtaining data to constrain the mathematical models. Instrumentation has been developed for measuring the velocities of grains in flight [24] and the flux of kinetic energy [20] and non-intruding measuring devices could be constructed based on laser-technology. An example of the latter possibility is Nickling's [38] experiments. These possibilities have not yet been exploited systematically to obtain constraining data.

3 Collision, reptation, creep

By the early 1980's, saltation calculations had been published by several authors, e.g. Owen [39] and White and Schulz [58]. The mechanics of sequences of saltation (by the same grain) had been explored by Tsuchiya and Kawata [51]. It was generally accepted that the other forms of grain motion, suspension and creep, were dependent on dislodgement effected by the collisions of saltating grains with the bed. The nature of these collisions was understood to be critical in the momentum transformations necessary for repeated saltation, and in determining, the number, size distribution and launch conditions of dislodged grains. Therefore considerable interest has been shown in grain collision and bed response since 1980.

At the 1985 Aarhus Workshop a general description of the collision process was given on the basis of direct observation via high speed film (Willetts and Rice [59]). The findings were endorsed by independent inferences of the outcome of collision obtained by calculation from multiple exposure photographs of the higher part of each saltation trajectory (Nalpanis [36]). Also in 1985, Rumpel [46] published a study which explored the influence of the bed geometry in the vicinity of the approach path of the incident grain on the rebound velocity vector. The grains in this study were spherical so as to make the geometry tractable, and no attempt was made to replicate the spatial randomness of bed grains in repose. Moreover, the bed grains were assumed fixed, so no displacement of these grains could take place in response to the collision.

Thus, by 1985, a good deal was known about the behaviour of the incident grain at collision, and also something about the ejections occasioned by the impact. However, very much less was understood about the rearrangement by collision of bed grains and the size distribution of the ejecta (including dust-size particles). The purpose of the present section is to give an account of some important developments in thought about near-bed grain activity which have taken place since 1985.

Collision has been studied since 1985 by direct observation and by numerical simulation. Willetts and Rice [60], [61], [63] have extended their direct observations of collision to investigate the influence of particle shape and of adverse bed slope. Grain shape was shown to have a pronounced effect on collision. Decreased sphericity is associated with lower velocities, shallower impact angles, and a lower degree of bed disturbance (fewer ejecta per collision). Direct dislodgement by wind forces was found to retain importance in a greater range of conditions in the case of less spherical, more platy, grains.

A local change of bed slope causes a change in the angle of approach to impact for the same inclination of the grain path to the horizontal. This is of interest principally in the study of ripples, which present the sand surface at a variety of slopes, up to $+13$ degrees on the stoss face and down to -30 degrees on the lee face. Change of angle of approach was found to affect the outcome of collision in several ways. With increased angle of incidence the ratio of ricochet speed to incident speed decreases and the ricochet angle increases. The vertical component of ricochet velocity increases with increasingly adverse bed slope. The forward momentum apparently lost at collision (as calculated from the observed momentum of the incident grain and of those disturbed grains which move clear of the bed) increases with increased angle of incidence and this, it was suggested, may be associated with increased creep activity. Here "creep" denotes motion which is undetected when the bed is viewed in side elevation. A distinction is implied between reptation which can be seen in that view and creep which can not. No pronounced change with incident angle was observed of the emergent speed of dislodged grains or their take-off angle relative to the bed.

Mitha et al. [34] also made direct observations of collision (using 4 mm steel spheres in the bed and for the impacting particles) as did Werner [55]. Mitha et al. confirm that the incident grain usually ricochets and that ejecta (of order 10 in number at each collision) have speed of emergence an order of magnitude lower than the incident speed. They report negligible correlation between ejection speed and either of incident speed or incident angle.

Very interesting theoretical studies of collision in two dimensions have been reported by Werner and Haff [56], [57]. See also Anderson and Haff [7]. A single-size population of circular "discs" are allowed to fall into a rectangular space and jostle elastically into stationary positions. The surface of the array so formed is then struck by an incident grain arriving at an appropriate oblique angle. The consequences of the collision include renewed jostling between the discs which results in changed stationary positions for many of them, usually rebound of the incident one, and ejection of a small number of surface ones. Despite the restrictions of this numerical calculation to two dimensions and to single size populations, the results are strikingly similar to data from direct observation of real collisions. The extension of the calculation to mixed grain sizes and three dimensions presents many difficulties but is now being embarked upon.

The importance of this type of study as a contribution to better understanding of impact lies in its potential flexibility with regard to incident grain velocity and the degrees of disturbance of the bed. Real observations are only practicable for very restricted ranges of approach speed and bed activity. Self regulating models of the saltation layer are now being constructed (see Section 5) and are nearing the point at which they can be used to study bed processes. Such models demand more secure splash laws over a wider range of u_* values than we have at present. Probably a combination of further observations and more numerical simulation will be necessary to make good the deficit.

Both the direct observations and the numerical studies of collision give only two dimensional data. The third dimension is not observed in the first case and does not exist in the second. This may be an important omission with regard to the bed microtopography (stationary grain arrangement) created by successive collisions. Such micro-topography may prove to have an important influence on collision. Image analysis and new spatial statistical methods are being used to explore the topography of the saltation affected bed (Barndorff-Nielsen [10]).

There have been few studies of reptation and creep. The majority of grains in motion travel in one of these modes although they are not dominant in determining transport rate because their mean excursion length is much shorter than that of saltating grains. Anderson and Hallet [8] and Sørensen [47] argue that the probability distribution of lift-off speeds is exponential, the lowest speeds being most probable. Excursion length is correlated sufficiently strongly with lift-off velocity, that its distribution would be expected to be broadly similar to the exponential function. Anderson [2] and Willetts and Rice [62] have used assumed reptation excursion length distributions in calculations of ripple growth and in interpretations of data concerning grain dislodgement rate, respectively. The tracer experiments of Barndorff-Nielsen et al. [11] have also been used by Sørensen [48] to derive a distribution of reptation excursion length. This may be distorted, however, by the absence of a detection system in the original experiment for short hops and fleeting rest periods.

It is clear that a considerable effort would be timely to examine carefully the grain movements which occur close to, and in, the bed. The extreme difficulty of direct observation must be overcome if the interesting postulations, now being incorporated in numerical models of the wind erosion process, are to be validated. They must be validated if a secure theory is to be achieved which will account for important features like surface sorting, the development of the ripple pattern, dust release, and soil deterioration.

4 Initiation of motion

Whereas the dislodgement necessary to sustain grain transport once this is established is generally believed to result from collision of saltating grains, the initiation of the process requires grains to be entrained by wind forces. This occurs when the wind strength rises to the fluid threshold and also when air blowing at greater than threshold speed over an immobile surface encounters the leading edge of a deposit of mobile material. Direct dislodgement by wind forces may also retain a role in established grain transport in some circumstances, but the significance of this role is not yet understood. This discussion will therefore be confined to initiation of motion by sand-free wind.

It has been usual to accept that the fluid threshold (the stage in a rising wind at which motion begins) can be characterised, for a particular sand, by means of u_*, the shear velocity. For example, Bagnold [9] suggested that the critical shear velocity can be calculated from

$$u_{*t} = A \left(\frac{\varrho_s - \varrho}{\varrho} \, gD \right)^{1/2} \tag{4.1}$$

in which ϱ_s and ϱ are the densities of grains and air, g is gravitational acceleration, D the equivalent grain diameter, A an empirical coefficient which depends on the friction Reynolds number, $u_* D / \nu$, and ν is the kinematic viscosity of air.

There have been two rival concepts of the manner in which grains enter saltation. Several authors, including Bagnold [9], describe grains as rolling and bouncing for perceptible distances before attaining the momentum to execute a saltation. Others, including Bisal and Nielsen [13], report that grains vibrate about a stationary position before being plucked by the airstream directly into saltation. However, several advocates of the latter view have worked with soil, rather than sand and thus with rather different restraining forces. Work reported in the last five years significantly clarifies the mechanism of direct dislodgement for sand grain sand therefore clears the way for a better prediction method for threshold conditions. This can be viewed as a better understanding of the dependencies of coefficient A in equation (4.1).

Nickling [38] made observations of the number of grains which moved sufficiently to interrupt a laser beam approximately 1 mm above the sand surface in a wind tunnel. He plotted the number of such interruptions per metre of beam per second against u_* and did the experiment for several sand grain sizes and glass bead sizes.

His method enables very specific judgements to be made about the threshold value of u_* which previously had depended on subjective assessments that "an appreciable number of grains" were in motion. The central finding is that general grain motion seems to cascade rapidly as those grains most susceptible to direct dislodgement collide with and disturb less susceptible grains downwind. The rapidity of this dislodgement depends on grain size, shape, sorting, and packing. There is somewhat tentative support for the view that directly dislodged grains enter saltation immediately, rather than after an interval of bouncing and rolling.

Williams [65] experimented on grain dislodgement near the leading edge of a flat plate set edge-on to an approach flow of uniform velocity. This arrangement enabled him to calculate the boundary layer condition at every point on the surface (assuming that grain motion does not affect this condition).

The fluid threshold condition was found to depend on the turbulence structure in the boundary layer. The dislodgement of a grain depends not only on its susceptibility, as postulated by Nickling, but also on the peak near-bed air velocities. In a flow which is not fully developed, peak values of velocity are not uniquely related to u_*, although this may be a permissible assumption in uniform conditions. The concept of Grass [22], originally devised to relate the probability distributions of grain susceptibility and of instantaneous fluid force in water, was invoked to demonstrate clearly the stochastic nature of direct dislodgement by air.

Williams found that the first translational movement of a grain, which may be preceded by a period of oscillation, can be either by rolling, or by ejection into saltation. Both sorts of initial motion occur. However, one of his most interesting findings is that initial disturbance is related to semi-organised "flurries" of activity which are often followed by relatively quiescent intervals. The flurries are attributed to the sweeps of flow associated with the well-known burst and sweep sequences in boundary layers (Kline et al. [31], Grass [23]). Grains dislodged in such flurries are likely to become the agents of further dislodgements by collision, as outlined by Nickling [38].

The picture which emerges of the near-threshold process is a complex one, even in the wind-tunnel where uniform air flows can be imposed. At scattered locations (but probably organised with respect to streamwise low speed streaks) almost random near-bed turbulence features seed the flow with low energy ejected grains. Many of these grains translate downwind at a range of speeds, dislodging other grains as they go. A single "flurry" therefore may give rise to a translating and dispersing sequence of dislodgements. Observations made at a particular station in near threshold flow involve the superimposition of several or many such dislodgement sequences.

One of the more daunting implications of this demonstration of the significance of turbulent velocity components at threshold of motion is that wind tunnel experiments are unlikely to give results of direct applicability in the field. The principal reason for this is that the near-bed layer in the atmosphere is usually in process of adaptation to large scale atmospheric disturbances or to topographic irregularity so that its turbulence is not that characteristic of a uniform steady flow having the same value of u_*. In particular, the distribution of bursts and sweeps in space, time and scale can be expected to differ from the relatively orderly pattern found in many wind tunnels. For this and for other reasons concerning the scale of turbulence, the characterisation of field-scale flows needs to go beyond a mean value of u_*. We do not yet have a compact extension of the flow description which can be justified in the context of sand transport threshold. Establishment of such a descriptor would be a most appropriate research target, not only for reconciling wind tunnel and field results, but also for reconciling the results obtained in different wind tunnels.

5 Self-regulating cloud models of aeolian saltation

In this section models for self-regulating saltation clouds are reviewed. By this is meant models of a saltation system into which are built a feedback mechanism that controls the amount of sand in transit. Such feedback mechanisms were discussed briefly in Section 2. Models of this type allow the following questions to be examined. What is it that determines the total flux in a saltating system, giving rise to the often documented nonlinear relation between transport rate and shear velocity? Why is there a hysteresis in this relation, which led Bagnold [9] to define two threshold velocities, the fluid and impact thresholds? What is it that determines the probability distribution of lift-off velocities, and the total number of grains in saltation given a wind of a certain speed? Why does the velocity profile during saltation depart from the law of the wall such that there exists a strong dependence of the perceived roughness of the bed on the driving stress, characterised by the shear velocity? Finally, how long does it take for the saltation system to respond to changes in driving stresses — information that constitutes a necessary prerequisite to the determination of long term fluxes resulting from realistically gusty wind histories.

Prior to the 1980's the principal work to address the saltation system in its full complexity was that of Owen [39], [40]. Owen noted the strong influence the saltation process exerted on the near-bed wind profile, and collected published data to reveal the functional dependence of the apparent bed roughness, z_0, on the imposed shear velocity u_* during sediment transport: $z_0 = \alpha u_*^2/2g$, where $\alpha = 0.02$. Owen saw this as evidence that the saltation layer thickness, which should scale as $u_*^2/2g$, in some way determined the effective roughness of the bed. In his theoretical development, Owen assumed that the nature of the feedback was that at steady state the shear stress exerted by the wind on the bed was reduced to that necessary to keep the surface grains in "a mobile state". He took this shear stress to equal Bagnold's impact threshold. For analytic simplicity, Owen assumed that all grains travel identical trajectories, and are launched vertically. In his calculation of the wind modification feedback caused by the saltating grains, he postulated a closure in which the turbulent structure was dominated by the wakes cast by the grains. Therefore the eddy viscosity was assumed constant, rather than proportional to the height above the bed. The latter assumption results in the law of the wall in the absence of sediment transport. Owen's eddy viscosity implies a strong deviation from a logarithmic profile in the saltation layer. The no slip condition at the bed was left unsatisfied.

The vigorous progress in our knowledge about the saltation process in the 1970's and 1980's has enabled the development of more complicated self-regulating models based on the new insights, and fast-growing computational efficiencies have allowed the relaxation of several of the important restrictive assumptions Owen made for analytic tractability. Work of White and Schulz [58] demonstrated that grains do not leave the bed vertically, and that in fact there is a wide spread of grain trajectories. Anderson and Hallet [8], while not addressing the physical reason for this spread in initial velocities, showed that many of the salient features of profiles of mass flux and concentration could be reproduced when such a broad distribution of initial conditions was incorporated. Sørensen [47] and Jensen and Sørensen [29], working with the extensive data sets collected by Williams [64], was able to constrain many of the important statistical properties of the saltation system from measured mass flux profiles. Based on their results they questioned that the feedback is as Owen described. Sørensen [47] was the first work to attempt to determine the magnitude of the wind feedback involved in aeolian saltation. The setting, then, in 1986, was one in which measured mass flux profiles had been understood in the light of the broad spread of grain trajectories that characterise saltation in air. The studies had not, however, addressed the issue of how this spread of trajectories came about.

Much of the important work since has therefore focused on the physics of the collision process (see Section 3), and on calculations involving both this and the wind feedback. The goals of such calculations are to reproduce the observed spread of grain trajectories and the observed profiles of mass flux, and to identify the nature of the wind modification feedback. While most of these developments have addressed the steady state saltation problem, most recent results allow treatment of the time-varying evolution of the saltation system.

Ungar and Haff [52] were first to publish a model that explicitly incorporated both the collision process and the wind field feedback. They sought a steady state solution wherein the grains ejected from the bed were both equal in number, and left the bed with the same velocity, as the preceding set of trajectories. The wind feedback was incorporated by defining a body force (per unit volume) on the wind operating in the horizontal (upwind) direction, which must be added to the Reynolds equation for the force balance on a parcel of air. This vertical profile of the saltation-imposed body force was calculated by summing the drag forces imparted to the grains in the system at each level above the bed. Steady state was achieved only when the splash process became stationary, i.e. when the lone grain trajectory in the calculation exactly reproduced itself upon impact. Their results were both intriguing and controversial. They suggested that in this simple model the grain trajectory does not depend upon u_*, and that the steady state transport rate should go as $u_*^2 - u_{*s}^2$, where u_{*s} is the shear velocity calculated from the air borne shear stress at the surface, rather than as u_*^3 as in numerous other studies (see Greeley and Iversen [25] for a list of such mass flux formulae).

Both Sørensen [47] and Anderson [1] calculated the wind field expected during steady state saltation by using empirically measured mass flux profiles to fix the numbers of grains in saltation and, in [47], the probability distributions of initial velocities of those grains. Their results differed because of differing assumptions made about the turbulent structure within the saltation layer. Sørensen assumed in closing the turbulent equation of motion that the eddy viscosity, K, remained scaled by the height above the bed, z, and by the far field shear velocity, u_*: i.e. $K = ku_*z$, where k is von Karman's constant. Anderson postulated that an "effective shear velocity" must be defined that depended upon the air borne shear stress at height z (see Section 2) rather than the shear stress in the grain free logarithmic layer. The effective shear velocity $u_{*s}(z)$ declined toward the bed from the far field u_* above the top of the saltation layer, so that the eddy viscosity $K = ku_{*s}(z)\,z$ decreased more rapidly toward the bed than Sørensen's did. The resulting steady state wind profiles differed primarily in that Anderson's showed considerably more curvature within the saltation layer — the influence of saltation was expected to be more pronounced. At the moment no measurements of the turbulence in the saltation layer have been published which can be used to verify the theories based on an eddy viscosity. However, these theories yield the proper shape of the modified wind velocity profile.

Werner [54], [55], in the most extensively documented model to date, has addressed the dependence of the steady state characteristics of the saltation cloud on the driving stresses. He extends the Ungar and Haff [52] treatment by incorporating a realistic splash law, which he develops through physical experiments based upon use of a sand grain gun. His results compare favourably with Ungar and Haff's conclusions in that the transport rate is seen to increase as $(u_*^2 - u_{*s}^2)^{1.2}$. The slight nonlinearity is attributed to the fact that the mean hop length is slightly dependent upon u_*, in contrast to Ungar and Haff's finding using the single trajectory model.

Anderson and Haff [6], [7] assembled all available information about the nature of the grain collision, and incorporated this into a model that simulated the saltation system from inception by aerodynamic entrainment to steady state. They assumed a simple model for the number of grains being injected into the air stream per unit area per unit time due to aerodynamic forces. This ejection rate was taken to be linearly proportional to the excess shear stress above the fluid threshold. The splash law combined information gleaned from both physical and numerical experiments. The wind feedback was as postulated by Ungar and Haff [52], and included the eddy viscosity closure used earlier by Anderson [1]. The results reproduced many of the salient details of the saltation system. The probability distribution of initial velocities at steady state was heavily skewed toward the low lift-off velocities resulting from grain splashes, resulting in the typically observed monotonic decline of mass flux and concentration profiles with height above the bed. The response time of the saltation system appeared to be on the order of $1-2$ seconds, or several long-trajectory hop times. The transport rate was shown to depend in a highly nonlinear way on the shear velocity (roughly a cubic power law), with a very rapid turn-on above the fluid threshold. In addition, the system displayed hysteresis when after steady state had been achieved the far field stresses were reduced to below the fluid threshold stress. New steady state fluxes could be achieved until the far field stress was made to fall below approximately $0.7-0.8$ of the fluid threshold, reproducing Bagnold's lower impact threshold.

It should be stressed that in these calculations a system with a number of incorporated forces was allowed to go where it wished. The results imply strongly that the nature of the feedback is not in restricting the number of aerodynamically entrained grains. Indeed these ejections vanished entirely before steady state was reached, as the wind field was modified substantially enough to reduce the fluid stresses at the bed to below those necessary to entrain grains. The feedback appears to lie, rather, in a combination of the wind modification and the grain splash process [4] wherein as the wind speed is diminished at all heights, the grains travelling through this field are accelerated less, reducing their impact speeds, which in turn, through the splash function, results in fewer grains ejected per impact. A balance is achieved only when on average a grain reproduces itself in the splash process, as initially suggested by Ungar and Haff [52] in their simpler model. Recent self-regulating saltation models are reported in [32] and [49].

In conclusion, while progress has been great in the last 5 years, in that the qualitative behaviour of the wind-driven saltation system is now largely understood, there remains much to be done. Theory has in many ways caught up to experiment, and is in need of new experimental work for verification of various postulated effects. For instance, the several-second response time implied by some of this modelling is difficult to test experimentally, as the wind tunnels used often have resonance times of this order. Yet the theoretical results to date have treated the most simple of systems. In the final section, we address the expected nature of the next generation of models and of problems that must be tackled by the aeolian community at large.

6 The link between saltation and suspension

The production and dispersion of suspended dust is a problem of considerable importance for the preservation of agricultural areas. Traditionally, the dispersion of suspended material has been described by a continuum model whose physical basis is rather less sound than that of the current saltation models. In recent years it has become clear, however, that suspension can be modelled in a way parallel to that of saltation. At least this is so in principle, but such an approach still poses a number of unsolved problems.

The traditional model for the variation of the volumetric concentration $c(x, z)$ of suspended fine particles, as a function of a horizontal coordinate x in the average wind direction and a vertical coordinate z, is the diffusion equation

$$U(x, z)\, \frac{\partial c}{\partial x} = \frac{\partial}{\partial z} \left(K(x, z)\, \frac{\partial c}{\partial z} \right) + w_s\, \frac{\partial c}{\partial z}, \tag{6.1}$$

see e.g. [19]. Here $U(x, z)$ is the time averaged wind speed at height z, $K(x, z)$ is the diffusion coefficient for the sediment, and w_s is the concentration-weighted average settling velocity of the grains. Usually horizontal homogeneity is assumed so that the left hand side of (6.1) is zero and the equation simplifies to

$$\frac{\partial}{\partial z} \left(K(z)\, \frac{\partial c}{\partial z} \right) = -w_s\, \frac{\partial c}{\partial z}, \tag{6.2}$$

an equation expressing the balance between gravitational settling and turbulent diffusion of the sediment. An expression for the diffusion coefficient $K(z)$ is obtained by setting it equal to the eddy viscosity of the air flow, which in a neutrally stratified atmosphere is given by ku_*z in the constant stress layer where the velocity profile is logarithmic. In the expression for the eddy viscosity u_* denotes the shear velocity, while k is von Karman's constant, 0.4. In the absence of sediment sources and sinks, (6.2) is solved to yield

$$c(z) = c(z_0)\, (z/z_0)^{-\gamma} \tag{6.3}$$

with $\gamma = w_s/ku_*$. Here z_0 is a reference level at which the concentration must be specified. The power law behaviour fits observed concentration profiles well. For an account, including several references, of the continuum model approach to aeolian suspension, see [8].

The continuum model approach still poses several interesting problems among which are the effects of a non-neutral stratification and of horizontal inhomogeneity. A very important problem is the proper specification of $c(z_0)$. Empirically derived formulae for the reference level concentration exist, see e.g. [8], but a more physically based solution would be theoretically more satisfactory and could be applied with more confidence. As will be discussed below, there is still a way to go before this goal is accomplished.

Anderson [3] calculated concentration profiles from computer-simulated ensembles of trajectories of small particles. The simulation took into account the random modifications of the grain path due to the vertical turbulent velocity fluctuations. The concentration profiles obtained in this way were well approximated by a power law profile, indicating that aeolian suspension can be modelled in the same way as modified saltation. In fact, suspension is nothing but an extreme kind of modified saltation. Models of this type would be physically more sound than the continuum theory, but also much more computationally intensive. A reasonable compromise, exploiting the fact that the two types of model give consistent results, would be to use the continuum theory far from the sand surface, but to set the concentration at a reference level close to the bed by means of a modified saltation model.

A coupling between saltation and suspension, where large high-energy impacting grains eject small grains from the bed, was already suggested by Bagnold [9]. The small grains can thus be mobilised at much lower wind speeds than required to lift them aerodynamically. Observations of aeolian transport of soils with a broad size distribution (Gillette et al. [17], [18], [21]) have shown conclusively that suspension of the fine grain fractions commences as soon as the first — generally larger — grains begin to saltate. As for pure saltation, a vital piece of information needed to understand the process of dust production is the splash law for multiple size grain beds; but, contrary to the situation when studying pure saltation, the interest is here focussed on the response of the fine grain fraction to impacts. The splash law for a grain bed of mixed sizes is a considerably more complicated object than the single size splash law.

The exact mechanism by which the fine grains are dislodged is not known. It might be a simple impact process as for the larger grains. It has, however, been proposed by P.R. Owen that fluid dynamical effects are important too. One possibility is convection of small grains in the wake of ascending saltating grains [40], another is aerodynamical lift induced by a vortex ring shed by an impinging saltating grain [41]. The lacking knowledge of the exact nature of the interaction between the fine grains in the bed and the saltating grains is an impediment to investigation of the splash law by means of computer simulations, because it is not obvious which, if any, aerodynamic mechanism to build in. This leaves experimental investigation the most appropriate way forward at the moment. This approach, however, faces severe problems concerning identification of the fine particles among the ejecta.

The splash law for the whole size-spectrum of ejecta together with the intensity of saltation in the equilibrium state would define the flux of fine grains from the bed into the air. Much of this dust settles again, so a model for the trajectories of fine grains is needed to enable calculation of the near-bed concentration profile. This profile would, in turn, define the reference level concentration needed to set the magnitude of $c(z)$ in (6.3).

Modelling of trajectories for fine grains whose motion is influenced by turbulent eddies was discussed by Hunt and Nalpanis [26], Nalpanis [36] and Anderson [3]. The two main problems are 1) to model the turbulent motion of an air parcel in the near-bed boundary layer and 2) to take into account the fact that grains do not follow the air flow exactly, but rather cut through the eddies of the turbulence. Given a realization of the wind history as seen by the moving grain, calculation of the grain trajectory is no more difficult than for pure saltation. In the equations of motion the time-averaged wind profile is simply replaced by the actual wind speeds experienced by the grain. These equations are easy to solve numerically, when lift forces are neglected. The problem is one of computer time: Since there are two sources of randomness — the launch velocity vector and the turbulence — an ensemble of many trajectory realizations, each with its own wind history, must be simulated for several launch velocities. Typically, one would like to repeat this for a number of size fractions.

Modelling of the motion of an air parcel in inhomogeneous turbulence, like for instance the atmospheric boundary layer, by means of stochastic differential equations has been a controversial subject. The problem is that effects of the non-linearity of the Navier-Stokes equation are not well represented by ordinary stochastic differential equations. However, it is possible to find reasonable approximate models of this type based on physical reasoning see [53] and [50]. Following the ideas of the former paper, Anderson [3] argued that in a neutrally stratified and grain-free atmosphere the vertical component w_t of the velocity of an air parcel can be adequately modelled by the stochastic differential equation

$$dw_t = -T_t^{-1} w_t \, dt + \varepsilon_t^{1/2} \, dW_t, \tag{6.4}$$

where

$$T_t = 0.4 z_t / u_*, \tag{6.5}$$

where dW_t is white noise (i.e. W_t is a Wiener process), and where

$$\varepsilon_t = u_*^3 / (k z_t) \tag{6.6}$$

is the turbulent dissipation at the air parcel position at time t. In (6.5) and (6.6) z_t denotes the distance between the air parcel and the surface. This model was also used by Hunt and Nalpanis [26].

The quantity T_t can be thought of as a so-called Lagrangian time scale, i.e. the time lag over which there is appreciable autocorrelation. Only in the case of homogeneous steady turbulence is T_t equal to the Lagrangian time scale in the rigorous sense, i.e. the integral of the autocorrelation function with respect to the time lag. Another problem is that the specifications (6.5) and (6.6) of T_t and ε_t are strictly valid only in the part of the atmospheric boundary layer where the velocity profile is logarithmic. In the saltation layer close to the bed it might be the case that the saltating grains tend to homogenise the turbulence so that the actual dependence of T_t and ε_t on the height above the surface is weaker than given by (6.5) and (6.6) in the presence of saltation.

The fact that the solid grains do not follow the air parcels, but rather move from one parcel to another, implies that the velocity fluctuations of the air as experienced by a grain are not modelled by (6.4) as it stands. It is intuitively obvious that the fluctuations seen by a grain must have a smaller autocorrelation than those of an air parcel. Therefore, a model for the wind history experienced by a moving grain can be obtained by shrinking T_t by a factor less than one, dependent on the slip between the grain and the air. For a discussion of ways to do this, see [2] and [26].

The main impediment to the approach to linking suspension and saltation outlined above is the lack of knowledge about the splash law for the whole size-spectrum of ejecta in multiple grain size beds. With this vital piece of information in hand, a first integrated model of suspension and saltation could, in principle, be constructed and used for computer simulations.

7 Future research directions

While the last five years have witnessed significant advances in our understanding of the complex set of processes that combine to produce sediment transport by wind, there remains an important and exciting set of problems to be dealt with in the future. Here we enumerate a few of these problems.

The experimental and theoretical research to date has focused on the simplest of systems, wherein the bed is assumed to be comprised of a single grain size, and the wind to be steady and uniform. The relaxation of these major assumptions, while difficult, is a prerequisite for addressing realistic geological and environmental problems.

7.1 The character of the bed

There is no hope of addressing such geologically interesting questions as the origin of small scale stratification, which is defined by textural variations in the involved sediments, without incorporation of multiple grain sizes in both experimental and theoretical work. In addition, the ejection of dust from mixed grain size beds is important in both agricultural settings and in environmental problems associated with the dispersal of fines from deflation of desert playa surfaces. This latter process has recently been recognized as figuring prominently in the formation of desert soils [15], and in the formation of major bauxite deposits [14], [35]. As discussed in Section 6, dust ejection has been known to occur at the inception of saltation, presumably because energetic saltation impacts blast these grains out of the bed mixture.

All of these problems require development of knowledge about grain collison in mixed grain size beds, and about the threedimensionality of that splash process. This poses two major problems for the theoreticians. First, calculation of flux profiles and of wind feedback during saltation of multiple grain sizes requires integration of a full range of trajectories for each grain size, meaning that calculation times will increase linearly with the number of grain sizes. Second, and more importantly, no longer may the bed be considered uniform either in the plane of the bed or in depth. Armour will develop, and will be spatially variable. The most obvious example of this spatial dependence is the ubiquitous coarsening of the crests of ripples.

In the past, saltation calculations have been performed by splitting the problem into two parts: first, a splash law was developed; second, a full saltation calculation was run wherein the splash process was assumed to be fully defined by the impact conditions of speed and angle. No spatial dependence was allowed. Nor was any temporal evolution of the character of the splash allowed. In abandoning the simple bed we are forced to account explicitly for spatial dependence by creating and allowing to evolve a spatially complex 2—3D bed over which saltation occurs. As the collision will depend upon the local grain size distribution in the bed, this will have to be known and tracked through the calculation. In addition, the (three-dimensional) spatial rearrangement of grains remaining in the bed will play a role in altering the nature of subsequent splashes. Saltated-upon beds "harden", reflecting the altered packing. The incorporation of an evolving bed will force by far the most radical departure of future models from present ones, and poses the most important challenge to the modeling community. While daunting, this next step should allow automatically the treatment of the associated problem of the evolution of mixed grain size ripples, and of the development of microstratigraphy.

The general problem of bed cohesion must be addressed, not only for the treatment of very small grains, but for instances in which moisture content plays a role. Clearly in the case of sand transport in coastal settings, the intergranular forces afforded by even quite low moisture contents are sufficient to alter greatly both the likelihood of aerodynamic entrainment, and the nature of the splash process. The mass transport may be controlled in large part by the rate of drying of the beach sands, rather than solely by the wind velocity. Careful field studies of coupled heat and sand transport are in order. This may perhaps best be done in coastal settings with moderate to strong diurnal (sea breeze) winds, see [27].

7.2 The character of the wind

The second major set of required advances lies in treatment of the complexity of the wind forcing. Relaxation of the steady-uniform approximation is absolutely essential in stepping into the real world. It has already been acknowledged (see Section 4) that the boundary layer burst and sweep mechanism plays an important role in the initiation of saltation. The problem becomes one of identifying the temporal and spatial scales of these first order variations in near-bed wind velocity, and just how these are affected by the presence of sediment transport. Are these, together with the temporal cascading effect of saltation [38], [6] responsible for the streaking of sediment transport, the sand and "snow snakes" characteristic of saltation in the real world?

It has been shown that the response time of the saltation system to changes in the wind velocity is of the order of a very few seconds. When combined with the highly nonlinear dependence of transport rate (and many other properties of the saltation system) on the shear velocity, this implies that in order to calculate well the total fluxes over long periods of time (minutes, hours, days) we must either know or be able to characterise the wind velocity history up to about 1 Hz frequencies. Seldom is such data collected; most meteorological stations collect wind velocities at 10 m heights averaged over periods of hours to a day. Significant effort should be expended in the search for an efficient manner in which to extrapolate from such data to realistic near-bed wind velocity, or better yet shear stress histories.

While the origin and evolution of aeolian ripples have recently been attributed to the spatial dependence of the bombardment intensity, while ignoring the effect of these small scale bumps on the air flow, any serious study of the origin and evolution of large aeolian bedforms (dunes, draas) requires a knowledge of the wind pattern, and in particular the shear stress pattern over such forms. While significant progress has been made in the understanding of these forms in the fluvial setting (e.g. [30], [16], [33], [37]), the scale of the required experiments in the aeolian setting has thus far defeated the acquisition of data necessary to confirm theoretical extension of the fluvial work to the aeolian setting. The numerical work currently being pioneered in threedimensional flow patterns over dunelike forms embedded in the planetary boundary layer will no doubt take great leaps forward in the next generation.

In short, then, while progress has been great in the last half decade, much work remains if the aeolian community is going eventually to tackle the realistic problems of geological and environmental interest. The promise of increased computing power will enhance our ability to address the multiple grain size and complex flow problems. Yet the theory has in large part caught up with the experimental and field aspects of the aeolian sediment transport field. There must in the next generation be a greater symbiosis between experimentalist and theoretician, both at the grain and at the bedform scales.

Acknowledgements

The research of R. S. Anderson was sponsored by the U. S. Army Research Office under grant number DAAL03-89-K-0089, and by the National Science Foundation (EAR-89-16102).

References

[1] Anderson, R. S.: Sediment transport by wind: saltation, suspension, erosion and ripples. Ph. D. Thesis, University of Washington (1986).

[2] Anderson, R. S.: A theoretical model for aeolian impact ripples. Sedimentology **34**, 943—956 (1987).

[3] Anderson, R. S.: Eolian sediment transport as a stochastic process: the effects of a fluctuating wind on particle trajectories. J. Geol. **95**, 497—512 (1987).

[4] Anderson, R. S.: The nature of the wind feedback in eolian saltation. Eos **69**, 1195 (1988).

[5] Anderson, R. S.: Saltation of sand: a qualitative review with biological analogy. Proc. Roy. Soc. Edinburgh **96B**, 149—165 (1990).

[6] Anderson, R. S., Haff, P. K.: Simulation of eolian saltation. Science **241**, 820—823 (1988).

[7] Anderson, R. S., Haff, P. K.: Wind modification and bed response during saltation of sand in air (this volume).

[8] Anderson, R. S., Hallet, B.: Sediment transport by wind: toward a general model. Geol. Soc. Am. Bull. **97**, 523—535 (1986).

[9] Bagnold, R. A.: The physics of blown sand and desert dunes. London: Methuen 1941.

[10] Barndorff-Nielsen, O. E.: Sorting, texture and structure. Proc. Roy. Soc. Edinburgh **96B**, 167—179 (1990).

[11] Barndorff-Nielsen, O. E., Jensen, J. L., Nielsen, H. L., Rasmussen, K. R., Sørensen, M.: Wind tunnel tracer studies of grain progress. In: Barndorff-Nielsen, O. E. et al. (eds.): Proceedings of the International Workshop on the Physics of Blown Sand, Memoirs No. 8, vol. 2. Dept. Theor. Statist., Aarhus Univ., Denmark, pp. 243—251 (1985).

[12] Barndorff-Nielsen, O. E., Møller, J. T., Rasmussen, K. R., Willetts, B. B.: Proceedings of the International Workshop on the Physics of Blown Sand. Memoirs No. 8. Dept. Theor. Statist., Aarhus Univ., Denmark 1985.

[13] Bisal, F., Nielsen, K. C.: Movement of soil particles in saltation. Can. J. Soil Sci. **42**, 81—86 (1962).

[14] Brimhall, G. H., Lewis, C. J., Ague, J. J., Dietrich, W. E., Hampel, J., Teague, T., Rix, P.: Metal enrichment in bauxites by deposition of chemically mature aeolian dust. Nature **333**, 819—824 (1988).

[15] Chadwick, O. A., Davis, J. O.: Soil-forming intervals caused by eolian sediment pulses in the Lahontan basin, northwestern Nevada. Geology **18**, 243—246 (1990).

[16] Fredsøe, J.: On the development of dunes in erodible channels. J. Fluid Mech. **64**, 1—16 (1974).

[17] Gillette, D. A., Blifford, D. A., Fenster, C. R.: Measurement of aerosol size distributions and vertical fluxes of aerosols on land subject to wind erosion. J. Appl. Met. **11**, 977—987 (1972).

[18] Gillette, D. A., Blifford, D. A., Fryrear, D. W.: The influence of wind velocity on the size distributions of aerosols generated by the wind erosion of soils. J. Geophys. Res. **79**, 4068—4075 (1974).

[19] Gillette, D. A., Goodwin, P. A.: Microscale transport of sand-sized soil aggregates eroded by wind. J. Geophys. Res. **79**, 4080—4084 (1974).

[20] Gillette, D. A., Stockton, P. H.: Mass, momentum and kinetic energy fluxes of saltating particles. In: Nickling, W. G. (ed.): Aeolian geomorphology. Boston: Allen and Unwin. pp. 35—56 (1986).

[21] Gillette, D. A., Walker, T. L.: Characteristics of airborne particles produced by wind erosion of sandy soil, high plains of West Texas. Soil Sci. **123**, 97—110 (1977).

[22] Grass, A. J.: Initial instability of fine bed sand. J. Hyd. Div. ASCE **96** (HY3), 619—732 (1970).

[23] Grass, A. J.: Structural features of turbulent flow over smooth and rough boundaries. J. Fluid Mech. **50**, 233—255 (1971).

[24] Greeley, R., Williams, S. H., Marshall, J. R.: Velocities of windblown particles in saltation: preliminary laboratory and field measurements. In: Brookfield, M. E., Ahlbrandt, T. S. (eds.): Eolian sediment and processes. Developments in sedimentology 38. Amsterdam: Elsevier 1983.

[25] Greeley, R., Iversen, J. D.: Wind as a geological process. Cambridge: Cambridge University Press 1985.

[26] Hunt, J. C. R., Nalpanis, P.: Saltating and suspended particles over flat and sloping surfaces. I. Modelling concepts. In: Barndorff-Nielsen, O. E. et al. (eds.): Proceedings of International Workshop on the Physics of Blown Sand, Memoirs No. 8, vol. 1. Dept. Theor. Statist., Aarhus Univ., Denmark, pp. 9—36 (1985).

[27] Hunter, R. E., Richmond, B. M.: Daily cycles in coastal dunes. Sed. Geol. **55**, 43—67 (1988).

[28] Jensen, J. L., Sørensen, M.: On the mathematical modelling of aeolian saltation. In: Sumer, B. M., Müller, A. (eds.): Mechanics of sediment transport. Rotterdam: Balkema, pp. 65—72 (1983).

[29] Jensen, J. L., Sørensen, M.: Estimation of some aeolian saltation transport parameters: a reanalysis of Williams' data. Sedimentology **33**, 547—558 (1986).

[30] Kennedy, J. F.: The mechanics of dunes and antidunes in erodible-bed channels. J. Fluid Mech. **16**, 521—544 (1963).

[31] Kline, S. J., Reynolds, W. C., Schraub, F. A., Rundstadler, P. W.: The structure of turbulent boundary layers. J. Fluid Mech. **30**, 741—773 (1967).

[32] McEwan, I. K., Willetts, B. B.: Numerical model of the saltation cloud (this volume).

[33] McLean, S. R., Smith, J. D.: A model for flow over two-dimensional bed forms. J. Hydr. Eng., ASCE **112**, 300—317 (1986).

[34] Mitha, S., Tran, M. Q., Werner, B. T., Haff, P. K.: The grain-bed impact process in aeolian saltation. Acta Mech. **63**, 267—278 (1986).

[35] Muhs, D. R., Bush, C. A., Stewart, K. C., Crittenden, R. C.: Geochemical evidence of Saharan dust parent material for soils developed on Quaternary limestones of Caribbean and Western Atlantic islands. Quaternary Res. **33**, 157—177 (1990).

[36] Nalpanis, P.: Saltating and suspended particles over flat and sloping surfaces. II. Experiments and numerical simulations. In: Barndorff-Nielsen, O. E. et al. (eds.): Proceedings of International Workshop on the Physics of Blown Sand, Memoirs No. 8, vol. 1. Dept. Theor. Statist., Aarhus Univ., Denmark, pp. 37—66 (1985).

[37] Nelson, J. M.: Mechanics of flow and sediment transport over nonuniform erodible beds. Ph. D. Thesis, University of Washington (1988).

[38] Nickling, W. G.: The initiation of particle movement by wind. Sedimentology **35**, 499—511 (1988).

[39] Owen, P. R.: Saltation of uniform grains in air. J. Fluid Mech. **20**, 225—242 (1964).

[40] Owen, P. R.: The physics of sand movement. Lecture Notes, Workshop on Physics of Desertification, Trieste 1980.

[41] Owen, P. R.: The erosion of dust by a turbulent wind. Lecture at the International Workshop on Sand Transport and Desertification in Arid Lands, Khartoum 1985.

[42] Rasmussen, K. R.: Flow over rough terrain. Proc. Roy. Soc. Edinburgh Ser. **96 B**, 129—147 (1990).

[43] Rasmussen, K. R., Mikkelsen, H. E.: Development of a boundary layer wind tunnel for aeolian studies. Geoskrifter **27**, Geologisk Institut, Aarhus Univ. (1988).

[44] Rasmussen, K. R., Mikkelsen, H. E.: The transport rate profile and the efficiency of sand traps. Preprint (1989).

[45] Rasmussen, K. R., Sørensen, M., Willetts, B. B.: Measurement of saltation and wind strength on beaches. In: Barndorff-Nielsen, O. E. et al. (eds.): Proceedings of International Workshop on the Physics of Blown Sand, Memoirs No. 8, vol. 2. Dept. Theor. Statist., Aarhus Univ., Denmark, pp. 301—325 (1985).

[46] Rumpel, D. A.: Successive aeolian saltation: studies of idealized collisions. Sedimentology **32**, 267—280 (1985).

[47] Sørensen, M.: Estimation of some aeolian saltation transport parameters from transport rate profiles. In: Barndorff-Nielsen, O. E. et al. (eds.): Proceedings of International Workshop on the Physics of Blown Sand, Memoirs No. 8, vol. 1. Dept. Theor. Statist., Aarhus Univ., Denmark, pp. 141—190 (1985).

[48] Sørensen, M.: Radioactive tracer studies of grain progress in aeolian sand transport. A statistical analysis. Research Report No. 141, Dept. Theor. Statist., Aarhus Univ. (1987).

[49] Sørensen, M.: An analytic model of wind-blown sand transport (this volume).

[50] Thomson, D. J.: Criteria for the selection of stochastic models of particle trajectories in turbulent flows. J. Fluid Mech. **180**, 529—556 (1987).

[51] Tsuchiya, Y., Kawata, Y.: Characteristics of saltation of grains by wind. Proceedings 13th International Coastal Engineering Conference, pp. 1617—1625 (1972).

[52] Ungar, J., Haff, P. K.: Steady state saltation in air. Sedimentology **34**, 289—299 (1987).

[53] van Dop, H., Nieuwstadt, F. T. M., Hunt, J. C. R.: Random walk models for particle displacements in inhomogeneous unsteady turbulent flows. Phys. Fluids **28**, 1639—1653 (1985).

[54] Werner, B. T.: A physical model of wind-blown sand transport. Ph. D. Thesis, California Institute of Technology (1987).

[55] Werner, B. T.: A steady-state model of wind-blown sand transport. J. Geol. 98, 1—17 (1990).

[56] Werner, B. T., Haff, P. K.: A simulation study of the low energy ejecta resulting from single impacts in aeolian saltation. In: Arndt, R. E. A. et al. (eds.): Advances in aerodynamics, fluid mechanics, and hydraulics. New York: American Society of Civil Engineers, pp. 337—345 (1986).

[57] Werner, B. T., Haff, P. K.: The impact process in aeolian saltation: two-dimensional simulations. Sedimentology 35, 189—196 (1988).

[58] White, B. R., Schulz, J. C.: Magnus effect in saltation. J. Fluid Mech. 81, 497—512 (1977).

[59] Willetts, B. B., Rice, M. A.: Inter-saltation collisions. In: Barndorff-Nielsen, O. E. et al. (eds.): Proceedings of the International Workshop on the Physics of Blown Sand, Memoirs No. 8, vol. 1. Dept. Theor. Statist. Aarhus Univ., Denmark, pp. 83—100 (1985).

[60] Willetts, B. B., Rice, M. A.: Collission in aeolian transport: the saltation/creep link. In: Nickling, W. G. (ed.): Aeolian geomorphology. Boston: Allen and Unwin, pp. 1—17 (1986).

[61] Willetts, B. B., Rice, M. A.: Collisions in aeolian saltation. Acta Mech. 63, 255—265 (1986).

[62] Willetts, B. B., Rice, M. A.: Particle dislodgement from a flat bed by wind. Earth Surf. Proc. Land Forms 13, 717—728 (1988).

[63] Willetts, B. B., Rice, M. A.: Collision of quartz grains with a sand bed: the influence of incident angle. Earth Surf. Proc. Land Forms 14, 719—730 (1989).

[64] Williams, G.: Some aspects of the eolian saltation load. Sedimentology 3, 257—287 (1964).

[65] Williams, J. J.: Aeolian entrainment thresholds in a developing boundary layer. Ph. D. Thesis, University of London (1986).

Authors' adresses: R. S. Anderson, University of California Santa Cruz, Earth Sciences Board, Applied Sciences Building, Santa Cruz, California 95064, U.S.A.; M. Sørensen, Department of Theoretical Statistics, Institute of Mathematics, University of Aarhus, DK-8000 Aarhus C, Denmark, and B. B. Willetts, Department of Engineering, University of Aberdeen, Kings College, Aberdeen AB9 2UE, Scotland.

Acta Mechanica (1991) [Suppl] 1: 21—51
© by Springer-Verlag 1991

Wind modification and bed response during saltation of sand in air

R. S. Anderson, Santa Cruz, California, and **P. K. Haff**, Durham, North Carolina, U.S.A.

Summary. A model of eolian sediment transport has been constructed, a special case of which is that corresponding to sand-sized mineral grains subjected to moderate winds: saltation. The model consists of four compartments corresponding to (1) aerodynamic entrainment, (2) grain trajectories, (3) grain-bed impacts, and (4) momentum extraction from the wind. Each sub-model encapsulates the physics of the process, and is constrained where necessary by experimental data. When combined, the full model allows simulation of eolian saltation from inception by aerodynamic entrainment to steady state.

Many observed characteristics of natural saltation systems are reproduced by the simulations. Steady state mass flux and concentration profiles all display rapid decay with height above the bed, representing the preponderance of short, low-energy trajectories in the saltation population. Yet the role of less abundant, longer, higher energy trajectories is a strong one: at steady state the entire population of saltating grains is controlled by high-energy bed impacts rather than aerodynamic entrainment. Because the nature of the grain splash process is such that high-energy impacts are much more efficient at ejecting other grains from the bed, the response time of the system to changes in wind velocity is determined by the hop time of these long trajectories. Several hop times, or roughly 1—2 seconds, are required.

Varying wind velocity among the simulation runs allows mapping of the relation between steady state mass flux and wind velocity — the mass flux "law" — which may be expressed as a power law of the excess shear velocity. The hysteresis that led Bagnold to define both fluid and impact thresholds for saltation is apparent, reinforcing our conclusion that it is the impacts of saltating grains that supports the large population of saltating grains at steady state.

1 Introduction

Considerable progress has been made in the understanding of the process by which sand sized grains are transported by the earth's atmosphere. Although ideas initially espoused by early workers, especially Bagnold [1] and Chepil [2], have enjoyed broad support over the decades through empirical findings, only recently has the theory of sand transport developed to provide the firm underpinnings to such empirical work. Theoretical work has now reached an approximate par with the empirical, and new questions raised by this theoretical work will require new field and wind tunnel efforts in the future.

The broad motivations for the study of the simple system of sand plus air range widely, but may be classed in four camps: environmental, geological, planetary and physical. The first motivated the work of Chepil and others, who lived in and through the Dust Bowl years of the US. Research founded by Chepil has been carried out since in extensive field and wind tunnel work centered at the Kansas research station, with the major emphasis being placed on the effects of wind on crops of the Great Plains and how to minimize these effects.

The geological motivations for performance of research on eolian sand systems include the need to understand the mechanisms by which eolian sandstone bodies of high porosity are produced, and how the grain packing geometry and other intrinsic variables of importance to the extraction of fluids from eolian rock reservoirs might differ according to the dominant variables in the problem: wind velocity, grain size distribution, etc.

The strong motivation to understand the settings into which we were sending planetary landing craft made the 1970's and early 1980's ripe for a resurgence of work on both the theoretical and empirical sides of the problem. Wind tunnels were constructed capable of very high and very low atmospheric pressures, in which both threshold velocities and mass flux relations were measured; high-speed filming techniques were used to track individual trajectories in saltation; interest was renewed in terrestrial settings in which analogues to Martian terrains were suggested by surface features; and theoretical work was renewed on both grain trajectories and threshold velocities for a wide range of atmospheric conditions.

The fourth motivation may be termed physical in that the research has been on the physics of the problem, rather than on its planetary, geological, or environmental applications. The several components of the problem have been taken apart and studied in detail: e.g., the grain trajectory, the grain-bed impact, the wind velocity feedback, and the aerodynamic threshold. Only recently have computational capabilities allowed putting all of these components back together again. In addition, interest was spurred by the degree to which the sand-air system was self-organizational, resulting in sand ripples of a well-defined wavelength.

2 The physical setting

It is useful to picture eolian saltation in its simplest manifestation as being comprised of four linked processes: aerodynamic entrainment, grain trajectories, grain-bed impacts (the granular splash problem), and the wind field modification. Clearly, no sand is transported at the very lowest wind velocities. Sand transport begins with increasing wind velocity as the most easily moved grains are displaced. The aerodynamics of the threshold condition must therefore be known, and quantified appropriately. Once airborne, grains travel trajectories that are determined by the balance between grain weight, drag and lift forces on the grain. The smaller the grains, the more likely they are to respond effectively to the turbulence of the boundary layer. We will consider in this paper only those grains whose size is large enough to filter out short time scale variations in wind velocity, meaning that they respond effectively only to the mean wind velocity profile: we therefore limit the discussion to saltation, rather than suspension (see Anderson [3]). Importantly, this reduces the grain trajectory problem to a deterministic one: once the liftoff conditions are specified, the entire grain trajectory is specified, including among other things the hop height, length, and hop time. Due to gravitational acceleration, all grains that are ejected from the bed ultimately return to the bed, upon which they interact energetically with a collection of other nearby particles. We must know in a quantitative sense the nature of this splash process, including what happens to the impacting grain, and to nearby grains. Because of the complicated microtopography of even a "flat" bed and because of the complexity of the packing geometry, this splash process is stochastic: we may know the results of any particular impact, characterized by the size of the impacting grain and its angle and speed, and the nature of the local bed (grain size distribution, angle with respect to horizontal), in only a statistical sense. All of the above-mentioned aspects of the saltation problem are dependent only upon the behavior of individual grains. The fourth facet to the saltation problem, that of the modification of the wind velocity, reflects the extraction of momentum from the wind by *all* of the particles in the air, and is an integral quantity that requires knowing the behavior of all of the airborne particles. In turn this modification will affect all of the other three processes, and may therefore be thought of as a negative feedback mechanism: aerodynamic entrainment will be curtailed to some extent, grain trajectories will be shortened by the fact that the entire wind velocity profile is slowed, and because the grain trajectories are slowed the impact velocities are diminished. Calculations such as we present here, which extend the work presented elsewhere (Anderson and Haff [4]) allow determination of the relative importance of these two feedback effects.

We will first introduce each of the four processes mentioned, with attention being paid in inverse proportion to attention in the present literature. We then present the results of calculations performed with all processes linked.

3 Aerodynamic entrainment

Grains may be entrained by direct aerodynamic forces of lift and drag. Aerodynamic entrainment is required for the initiation of saltation in the absence of external mechanical disturbances. As the simplest possible model for aerodynamic entrainment, we take the number of entrained grains per unit area of bed per unit time (the aerodynamic ejection rate) to be proportional to the excess shear stress

$$N_a = \zeta(\tau_a - \tau_c) \tag{1}$$

where τ_a is the short term mean shear stress at the bed, τ_c is the critical fluid shear stress for entrainment, for which an expression has been determined from numerous wind tunnel experiments, and ζ is a constant (here chosen to be of order 10^5 grains Newton^{-1} sec^{-1}). All such aerodynamically entrained grains are assumed to leave the bed with a velocity corresponding to that needed to reach a height of about one grain diameter, D, above the bed. This is a conservative assumption that effectively limits the role of the aerodynamically entrained grains in the eolian saltation process. As will be demonstrated below, this somewhat ad hoc entrainment rule does not play a critical role in a well-developed saltation curtain.

4 Grain trajectories

Once launched into the airstream by either ballistic impact forces or aerodynamic forces, grains are acted upon by the surface tractions of drag and lift forces imposed by the wind, and by the body force of the grain's weight. Many studies have addressed the details of the forces acting upon airborne grains, beginning with Bagnold [1]. They may be simply stated as follows. Consider a coordinate system in which positive x is in the downwind direction and positive z is upward, Fig. 1. The particle weight is $\boldsymbol{F_w}$, written

$$\boldsymbol{F_w} = M_p \boldsymbol{g} \tag{2}$$

where M_p is the particle mass and $\boldsymbol{g}$ the acceleration due to gravity. The drag force, $\boldsymbol{F_d}$, operates in the direction of the relative velocity of the grain and the local wind velocity,

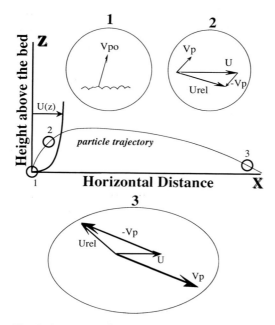

Fig. 1. Schematic diagram of coordinate system, X, Z, particle trajectory and mean wind velocity, $U(z)$. Particle launched at velocity V_{po} (site 1) is seen at two other positions within its trajectory. Relative velocity between particle and air is obtained by vector subtraction: $\boldsymbol{U}_{rel} = \boldsymbol{U} - \boldsymbol{V}_p$. High on upward trajectory (site 2), relative velocity is such as to reduce upward vertical velocity of the particle, while on the downward trajectory, it is such as to reduce downward vertical velocity. Magnitude of the wind velocity, U, increases with height above the bed

$\boldsymbol{U}_{rel}$, and may be written

$$\boldsymbol{F_d} = \frac{1}{2}\, \varrho_a C_d A \, |\boldsymbol{U}_{rel}|\, \boldsymbol{U}_{rel} \tag{3}$$

where ϱ_a is the density of air ($= 1.22$ kg/m³ at 20 °C), A is the cross-sectional area of the grain presented to the flow ($= \pi D^2/4$ for spheres, where D is grain diameter), and C_d is the drag coefficient, which is dependent upon the local Reynold's number, $\mathrm{Re} = U_{rel}D/\nu$, where ν is the kinematic viscosity of air ($= 1.5 \times 10_0^{-5}$ m²/s at 20 °C).

As the shear-imposed lift force on a spherical grain is essentially negligible at heights more than a few grain diameters above the bed (Anderson and Hallet [5]), we may neglect it in the first order sketch of the trajectory problem presented here. More elaborate and rigorous treatments will incorporate this effect as well as others, among them the Magnus effect (White and Schulz [6]) and the virtual mass effect. The essence of the trajectory problem is, however, contained in the weight and drag forces. The downstream and vertical components of these two forces may be added to yield the resultant forces in the x and z directions, from which accelerations, $\boldsymbol{a}$, velocities, $\boldsymbol{U}$, and displacements, $\boldsymbol{x}$, may be calculated for a given time step, dt, from Newton's second law:

$$\boldsymbol{a} = \frac{d\boldsymbol{U}}{dt} = \frac{\boldsymbol{F_w} + \boldsymbol{F_d}}{M_p}; \qquad \boldsymbol{U} = \frac{d\boldsymbol{x}}{dt} = \frac{d\boldsymbol{U}}{dt}\, dt; \qquad \boldsymbol{x} = \frac{d\boldsymbol{x}}{dt}\, dt. \tag{4}$$

Given the initial conditions of liftoff angle and speed, these equations can be integrated forward in time until the particle reimpacts the bed. We find that very accurate calculations of trajectories may be accomplished with 100 time steps per trajectory, requiring that the time step be chosen such that $dt = \tau_{hop}/100$, where τ_{hop} is the expected hop time in the absence of vertical drag, or $2w_0/g$, w_0 being the initial vertical velocity of the grain. [It is the happy case in the eolian trajectory problem that the accelerations in the vertical are due primarily not to fluid drag but to gravity, meaning that hop times and hop heights are well approximated (less about 10—20%) by their values in a vacuum (see Anderson and Hallet [5]).]

5 The splash process

Our knowledge of the splash process has been augmented greatly in the last decade by both physical and numerical experiments. The sophistication of these experiments has advanced considerably, and has resulted in placing us in a position to characterize in considerable detail the complexity of the process. Saltation is a stochastic process that must take into account the reaction not only of the grain impacting the bed, but the resulting rearrangement of the grains in the bed, including the possibility of ejecting some of these grains. The impact process is expected to depend upon the mass and velocity of the impacting grain, including its speed and approach angle with respect to the local bed slope, and the masses and elastic parameters describing the grains comprising the bed near the impact point.

Early numerical experiments were crude in the sense that they took into account only the geometrical complexity of the problem, while ignoring the dynamical nature of the problem. Grains in the bed were not allowed to be ejected, and the splash process was reduced to finding the distribution of rebound angles to be expected from beds of particular micro-topographies. Simple geometrical models such as these predict that if the bed were perfectly rigid the rebounding particle would most likely emerge from the collision at roughly 50° from the horizontal, having impacted the bed at angles typical of eolian saltation, i.e. 10—15°. This matches closely the measured averages of saltation angles in full saltation experiments, as first reported by White and Schulz [6], and more recently in numerous publications by Willetts and Rice (e.g., [7]—[10]). This was also the first work to question seriously the conclusions of Bagnold [1] that the preponderance of grains were ejected from the bed vertically, conclusions that had played an important role in shaping later theoretical work (e.g. Owen [11]).

By utilizing an elevated strip of bed and high-speed films, Willetts and Rice at the University of Aberdeen have developed a statistical description of the impact process during sparse saltation in wind tunnel settings. Typical saltation impacts strike the bed at angles of 10—15° from the horizontal. The impact results in two phenomena: the rebound of the impacting grain, and the emergence of several grains from among the bed grains. The rebound speed is reduced by roughly half from the pre-impact value, and the mean rebound angle is roughly 50° from the vertical. The number of splashed grains rises roughly linearly with the impact speed, being of the order of 3 grains per impact for impact speeds on the high end of those expected in eolian saltation. The ejection speeds of the splashed grains are of the order of 10 percent of the impact speed, and the ejection angles are canted more steeply toward the vertical than the rebound, within a broad range.

These measurements have been used to check the results of numerical simulations that attempt to take into account all the forces between all the grains in the problem. This was first addressed in detail by Werner (e.g., [12], [13]) in his dynamical simulations of the impact process. Working with well-packed two dimensional beds, impactor grains could be fired into the beds at any chosen angle and speed. The statistics from many such impacts were collected to describe the outcome from any set of impact parameters. Werner reported distributions of rebound and ejection angles that paint a picture similar to that obtained from the high speed films. Most of the grains ejected from the bed were found to emerge from within the top one to two grain layers, and displayed a high ejection angle. Werner noted that this was probably due to the fact that this was the direction of the free face, and therefore the direction in which there was least resistance to motion. He further noted that when the bed was given some micro-relief, that grains were preferentially ejected from micro-discontinuities he termed steps. This allowed a greater range of possible ejection angles, and therefore resulted in a greater spread of the resulting ejection angle distribution. Simultaneously, the Caltech group (e.g., Mitha et al. [14]) developed a means of producing and documenting the results of individual grain impacts using real particles. These experiments started using spherical steel pellets (bb's) which were fired at a given angle into a wide bed of similarly sized pellets. Subsequent development of a "sand gun" allowed similar experiments using individual sand grains fired into sand beds (Werner [12]).

Subsequent to Werner's simulations, Anderson and Haff [4] reported further dynamical simulations that attempt to map out the full splash function. The well-packed nature of the simulated bed is abandoned in these calculations. Several two-dimensional simulated beds are created by dropping grains from a given height with a random distribution of initial velocities. The particles are allowed to settle in a "box" that is constructed with wrap-around boundary conditions that effectively extend the bed laterally. The resulting beds manifest more realistic micro-topography that importantly affect the splash outcome. These beds are then each subjected to impacts with identical impact parameters of speed and angle, many impacts being performed on each of three beds, each impact impinging on the bed at a different point in order to collect sufficient statistics. The resulting splash descriptions therefore have embedded in them the stochastic nature of the bed, as well as the possibility of hitting any particular grain in the bed at any of many different points.

6 Impact model

The splash function used in these calculations is generated from direct computer simulation of the motion of the impacting and impacted grains. The simulation is performed in two-dimensions, with each grain assumed to be circular, Fig. 2. The position and velocity of each particle in the simulation is followed explicitly through time via integration of the Newtonian equations of motion

$$\boldsymbol{F}_i = M_i \boldsymbol{a}_i \tag{5}$$

where $\boldsymbol{F}_i$ is the force applied to particle i, M_i is the particle mass, and $\boldsymbol{a}_i$ is its acceleration. Since friction between particles is included in the model, torques may be applied leading to particle rotation. To the three degrees of freedom attaching to the motion of a two-dimensional extended particle, namely translational motion in the x and z directions, plus angular motion about the center of mass, there correspond the two components of Newton's second law, $F_x = Ma_x$ and $F_z = Ma_z$ plus a torque equation,

$$\tau = I\alpha \tag{6}$$

where τ is the applied torque, I the moment of inertia about the centre of mass, and α the particle's angular acceleration. Following the motion of each particle therefore requires the integration of 3 second order or 6 first order differential equations, for a total of $6\,N$

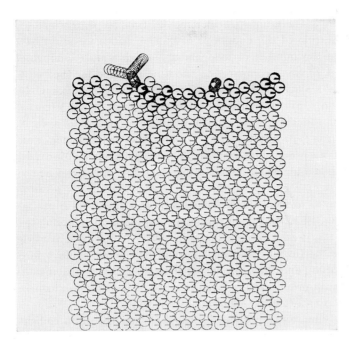

Fig. 2. Computer simulation of spherical grain impacting bed of 500 identical spheres (coming from left to right). Note realistic roughness of the surface, which determines the geometrical character of the rebound. Rebounding grain (seen here at 21 equally spaced time steps) scatters forward in this particular impact. Many other grains in the top few layers of the bed are affected, as evidenced by the fuzzing of their boundaries); at least one appears as though it will be ejected from the bed

equations for a model impact system with N grains. Typical impact scenarios have involved 80 to 100 grains, corresponding to $\sim 500-600$ simultaneous equations. There are two basic types of force which we must consider, body forces and contact forces. Contact forces can be categorized as normal (or compressive), and tangential (or frictional). The relevant body force here is just the weight of the grain, $\boldsymbol{Mg}$ (we drop the subscript i for now).

The essential physics to be captured by a contact force model includes the fact that sand grains are very "stiff", i.e., do not deform significantly upon contact, that a collision between two sand grains is not quite elastic, and that real sand grains are frictional. At the present time, modelling three-dimensional grains or grains of non-circular shape lies beyond our computational capabilities, a restriction which should begin to fade within the next several years.

The normal force is modeled as a linear spring. This force is not activated until two particles "contact" one another, as determined by the spacing of their centers becoming less than the sum of their radii. The stiffness of the spring is chosen great enough that particle "overlap" is always less than a predetermined fraction of the particle diameter, a consideration which depends upon the maximum relative particle velocities expected to occur in the simulation. In the saltation problem, this is always determined by the speed of the impacting grain. Higher speeds require stiffer springs. Since the natural period T of a linear spring decreases as the spring constant k increases,

$$T = 2\pi \sqrt{M/k} \tag{7}$$

an increase in k implies an increase in the number of incremental time steps Δt which are needed to effect the actual integration of the equations of motion (since we must have $\Delta t \ll T$). This is one limitation on the number of independent splash simulations which can in practice be run, i.e., T must be small compared to the time required for the bed to respond sufficiently that final grain trajectories can be determined. For a grain of diameter $D = 0.05$ cm, density $\varrho = 2.5$ g cm^{-3}, and spring constant $k = 2.0 \times 10^6$ dy cm^{-1} (corresponding to a maximum overlap of about 2% of the grain diameter at an impact velocity of 100 cm s^{-1}), $T = 100$ µs. And Δt is generally taken to be at least an order of magnitude smaller in order to ensure adequate accuracy of integration. (The integration itself is carried out using a predictor-corrector scheme for which preset integration tolerances can be specified.)

These numbers begin to illustrate some of the technical problems involved in particle dynamics modelling. If $\Delta t \sim 10^{-5}$ s, then 10^5 integration steps are required to simulate 1 second in the sand grain world. Since each integration step involves, for each particle, checking for contacts, evaluating the (usually multiple) normal and shear contact force components (and the resulting torque), and then computing the actual changes in velocity and displacement components and in orientation, a large number of floating point operations are required for even relatively short intervals of simulated time.

The calculation is made slightly more complicated by the fact that the grain collision is to be damped, i.e., the grains are slightly inelastic. Damping is modeled by a velocity-dependent force which is proportional to the instantaneous normal relative particle velocity and which always opposes relative particle motion. In such a damped oscillator, there are two time scales, one is the effective or modified spring frequency, which is less than the natural (undamped) frequency, and the other is the damping time scale which gives the e-folding time for degradation of the oscillator amplitude. For an accurate integration of the equations of motion, the integration time step Δt must be less than both of the system physical time scales.

For such an oscillator there are three characteristic response modes, namely, underdamping, critical damping and overdamping, depending upon whether the damping time-scale is greater than, equal to, or less than the natural frequency of the system. Critical and overdamped behavior correspond to collisions in which the two colliding partners lose all their normal velocity in a head-on collision (i.e., the coefficient of restitution, ε, given by the ratio of outgoing to incoming (normal) relative velocity, is zero). For the collision of resilient quartz grains the coefficient of restitution is greater than zero (it would be unity for a perfectly elastic collision) and consequently the normal forces in the impact simulation are modeled by underdamped springs. The choice of spring constant and damping strength is therefore made to satisfy the overlap requirements, and to produce a desired coefficient of restitution. The precise value of ε to be used in a given simulation is somewhat problematical, because, at the collision speeds characterizing high-speed impact events (a few ms^{-1}), the actual grain contact is not at all elastic, but rather is characterized by plastic deformation, spalling and chipping off of small pieces of the parent grains. Simulation runs on identical systems with reasonable variation of ε show that the main details of the splash function are not very sensitive to the exact value of ε. In our previous simulations we have taken $\varepsilon = 0.7$.

Besides the normal compressive and corresponding damping force, which act in a direction along the line of centers of two contacting spherical particles, the simulation model also includes a tangential force due to friction at the point of contact. Contact friction is modeled in one of two different ways, depending upon whether, at any instant, there is slippage at the contact or not. If the contact is slipping, i.e., if there is relative tangential motion of the material just on either side of the contact point, we invoke a rule corresponding to Coulomb friction, i.e., that the tangential frictional force is related to the instantaneous normal force F_n via a constant of proportionality, the friction coefficient μ. If the contact is *not* slipping, then we imagine that a tangential, damped, linear spring acts to oppose any slippage. Particles may still roll freely because the friction spring is a function of the total arc length separating the anchoring points of the spring minus the amount of rolling. If slippage starts to occur, the very stiff tangential spring effectively limits rim displacements unless or until the spring force exceeds μF_n. If this occurs, then the tangential spring is severed, and the maximum tangential force remains at μF_n. The tangential spring is damped in order to suppress low-amplitude, high-frequency (unphysical) tangential oscillations of the contacting particles.

The choice of a friction coefficient is, like that of the coefficient of restitution, somewhat problematical also, because, while perfectly smooth quartz grains exhibit low friction, real sand grains are pitted, faceted and generally irregular in shape. Thus, in nature, torques can be exerted across a particle contact even in the absence of true friction (the Clayhold effect). The friction coefficient μ, for the sphere model described here, is clearly an effective friction coefficient which incorporates not only the effects of true surface friction, but also effects of particle shape and surface morphology. In our saltation calculations, we have used a generic value of 0.5 for the friction coefficient. As with the coefficient of restitution, the splash function is only modestly sensitive to μ, as long as it is not too small.

Each saltation impact calculation is carried forward in time far enough that the nature of the bed-response is well-established. Depending upon bed conditions this might be 10 T —20 T. If the simulation is followed too far in time, complications can arise from the presence of stress waves reflected off the substratum upon which the simulated collection of bed-grains rests. These waves, which are artifacts of the calculation, can actually throw grains off the surface and lead to an overestimation of the number of reptating grains. This is one of several computational issues which arise as a result of the necessarily small number of grains which can be handled in each impact event. In our simulation studies the "target" bed was prepared by dropping grains onto a fixed and immobile surface. The grains were allowed to fall under the influence of gravity and to rebound back again, colliding with one another however they might, until friction and the inelasticity of collisions reduced the kinetic energy of the grain mass to essentially zero. Because of the relatively small number of grains, periodic boundary conditions were applied at the edges of the region of calculation, as in Fig. 2. At these lateral boundaries, particles and their contacts are "wrapped" around to the other side of the cell, so that if a particle moves too far to the right, it is inserted again on the left, at precisely the same elevation it had on the right-hand side. While periodic boundary conditions can introduce spurious behavior in certain situations (especially in particle flow studies), they seem to be relatively innocuous here because the calculation of every impact is wisely terminated before the disturbance initiated by the impact can propagate out one side of the test region and back in the other.

Once a bed has been constructed in this way, we impinge an identical projectile of given velocity and impact angle upon it. Ten impact positions were chosen for each velocity and impact angle combination, corresponding to equally spaced impact points for a reference flat bed. Since the bed has its own unique topography, the actual impact points were *not* uniformly distributed. Thus the "upwind" side of any clumping of grains at the surface will tend to receive somewhat more impacts than the "downwind" side of such a clump, as would be the case in nature. (This is the physical origin of wind-ripple instabilities.)

One technical issue of importance in grain simulation revolves around contact detection. Two grains are in contact whenever the separation between their centers of mass is less than the sum of their radii. Since all particles in the simulation may potentially move, we must have a method for periodically detecting the generation of new contacts and the breaking of old ones. One way to do this is, at every integration step, to check the separation of the centers of mass of every pair of particles at each integration step. If there are N particles in the system, then there are $N(N-1)/2$ pairs, so the number of pairs to be checked increases like N^2 for large N. Implementing contact detection this way leads to a rapid saturation of computing power when the number of particles becomes large. Various techniques exist for cutting down the fraction of computer time spent in contact detection. Most of them involve some sort of fine- or cross-graining and take advantage of the fact that a grain's neighbors (contacting or not) at time t will still be the grain's (only) neighbors at time $t + \Delta t_{\text{check}}$, where Δt_{check} is large compared with the integration time step Δt, but small compared with the total elapsed time in the simulated system, t_{total}. Thus, during an interval of time Δt_{check} it is necessary to check only $n \ll N$ contacts within a small neighborhood of each particle. Only occasionally (at Δt_{check} intervals) need all pairs be checked.

The eolian saltation impact program implements a variation of this method, in that each particle carries attached to it a "neighbor list", which contains an identifying particle number for every (other) particle that was within a specified radius at time t. At each Δt, the collision detection code looks only at the neighbor list. As old neighbors move away and new ones move in, the list is updated, using a rule based upon the speed of the fastest particle in the system.

Numerical simulations of single grain impacts into a granular bed were performed with the aim of providing a quantitative picture of the splash process. The impact simulations were performed in two dimensions with identical slightly inelastic grains (spheres) whose interactions were characterized by a coefficient of restitution $\varepsilon = 0.7$, an inter-grain friction coefficient $\mu = 0.5$, and a contact spring constant $k_{\text{eff}} \sim 10^7$ dynes cm^{-1}. Beds were

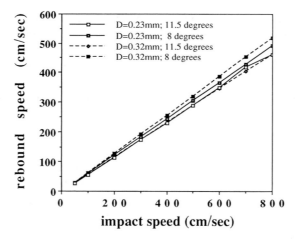

Fig. 3. Mean rebound speed as a function of impact speed, from computer simulations of grain impacts. Each point represents the mean of 20 impacts into a bed of 500 grains, at a fixed angle. Four cases are shown: little grains ($D = 0.23$ mm) impacting beds comprised of similar grains, at 8 and 11.5°; and big grains ($D = 0.32$ mm) impacting a bed of similar grains a 8 and 11.5°. Mean rebound speed in each case is approximately 50—60% of impact speed. The influence of impact angle is slight, in each case resulting in slightly lower rebound speeds for higher impact angles

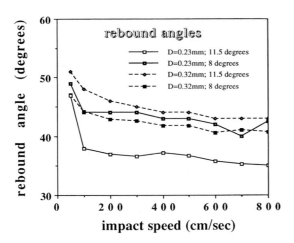

Fig. 4. Mean rebound angle as function of impact speed for same cases presented in Fig. 3. Mean rebound angle is approximately 35—45° from horizontal, much higher than the incidence angle. For the larger grains, higher angles result in higher mean rebound angles, by roughly the difference in the impact angle. For the smaller grains, higher impact angles result in lower mean rebound angles

generated by dropping 87 particles with random initial velocities into a box with periodic boundaries, its width chosen such that the resulting bed was about 10 grains deep. Three such granular beds, distinguished by different packing configurations, were each impacted 20 times to develop the splash statistics for each impact angle and speed. These computations soften assumptions made in earlier numerical work on impact dynamics in which the bed was assumed to be hexagonally packed (Werner [12]). Impact angles and speeds were chosen to cover the ranges typical of eolian saltation impacts in air (angles 8—15°; speeds 25—800 cm/sec). While earlier simulations (Anderson and Haff [4]) used grains of diameter $D = 1$ mm, more recent calculations have been performed using more typical eolian sand sizes: $D = 0.23, 0.32$ mm.

The picture of the grain splash resulting from our numerical simulations is qualitatively similar to that derived from earlier numerical (Werner [12], [13]) and physical experiments (Mitha et al. [14]; Willetts and Rice [10]) and fills in gaps in these latter results, especially in the low-impact velocity range. In our simulations the impacting grains were found to rebound from the surface with nearly unit (0.95) probability for the majority of the impact speeds simulated, a result postulated earlier by Rumpel [15]. For low impact speeds the probability of rebound can be expressed as

$$P_r = 0.95 \left(1 - \exp\left(-\xi V_{\mathrm{im}}\right)\right), \tag{8}$$

where ξ represents the inverse of a velocity scale for the approach of the probability to its impact velocity asymptote. We find $\xi = 2.0$; i.e., for $V_{\mathrm{im}} = 0.5$ m/s the probability of rebound is roughly 0.7. In the rare cases where the incident grain fails to rebound at high impact speeds, the micro-geometry of the collision pocket is such that the momentum of the impact is efficiently delivered to one or more nearby grains, while the colliding particle is geometrically trapped.

For the chosen grain parameters, the mean rebound speed is approximately 50—60% of the impact speed, Fig. 3, and the mean rebound angle is 35—45° from the horizontal, Fig. 4. The ejection angle is therefore significantly steeper than the impact angle, as suggested by earlier work of Rumpel [15], Werner and Haff [16], [17], and others.

The impact forces result in ejection of a number of grains from within a few grain diameters of the impact site. While there exists a large range of bed response for a fixed set of impact conditions, Fig. 5a, the mean number of grains ejected increases roughly linearly with impact speed, and compares well with data obtained from physical impact experiments with coarse sand, Fig. 5b. The mean speed of the ejected particles is considerably lower than that of the rebounding grain, and is roughly 10% of the speed of the impacting grain, for reasonable impact speeds, Fig. 6. The mean ejection angle tends to be oriented downwind

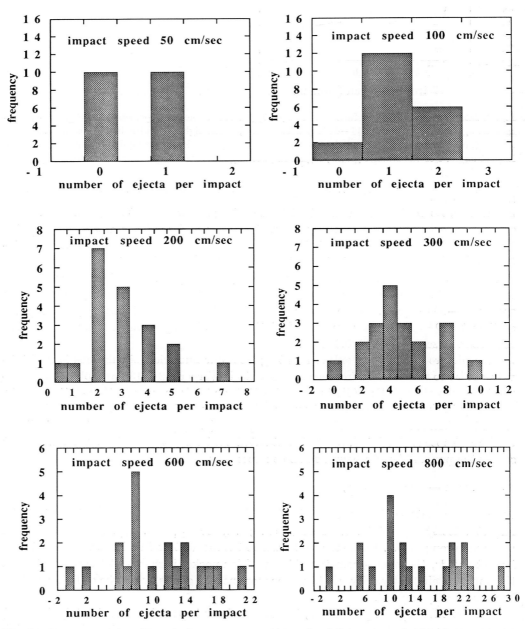

Fig. 5a. Histograms of number of ejecta per impact for each of 6 impact speeds, showing large range of bed response. Diameter 0.23 mm impactor into bed of similar grains at 11.5°. Total number of impacts for each case = 20

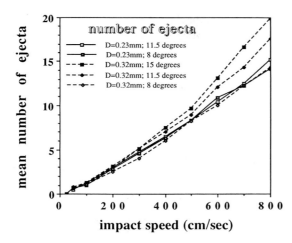

Fig. 5b. Mean number of ejecta per impact as a function of impact speed for 5 cases: 8 and 11.5° for $D = 0.23$ mm grains impacting beds of similar grains; 8, 11.5, 15° for $D = 0.32$ mm grains impacting beds of similar grains. Mean number of ejecta increases linearly with impact speed; these results compare well with data obtained from physical impact experiments (e.g. Werner [12], [13]; Willetts and Rice [7], [8], [9])

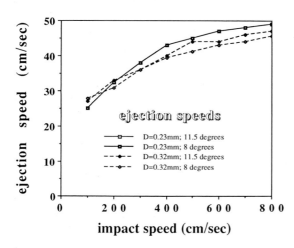

Fig. 6. Mean ejection speed as a function of impact speed for each of 5 cases shown in Fig. 5. Results for 8 and 11.5° cases with small grains ($D = 0.23$ mm) exactly overlap. Mean ejection speed appears to increase less than linearly with impact speed, at roughly $(V_{im})^{1/2}$. For typical or intermediate impact speeds of several m/s, mean ejection speed is $\sim 10\%$ of speed of impacting grain

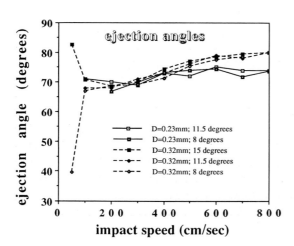

Fig. 7. Mean ejection angle tends to be oriented downwind $\sim 70°$ from horizontal for all 5 cases, and is roughly constant across a large range of impact speeds

at 60—70° from the horizontal, Fig. 7. All results for impacts between 8° and 15°, which cover the range of expected impact angles in eolian saltation, display only slight dependence on the impact angle. These results are expressed in terms of the splash function (Ungar and Haff [18]), which describes the number density of grains ejected from the bed as a function of their liftoff velocity, for a given impact velocity.

7 Discussion of the single impact simulations

The picture of the grain splash emerging from the numerical experiments is both qualitatively and quantitatively similar to that arising from physical experiments (e.g., Mitha et al. [14]; Willetts and Rice [7], [8], [9]), and from earlier numerical simulations (e.g., Werner and Haff [16], [19]). A most impressive correspondence comes with the data on impacts collected with high speed motion photography taken during saltation in a wind tunnel (Willetts and Rice [7], [9]).

Although considerable caution is warranted in relying upon these numerical experiments, which involve two-dimensional, perfect circles, the correspondence with physical experiments using real sand grains in three-dimensional pockets is encouraging. The low impact velocity results are the most uncertain, since experimental confirmation is difficult due to problems associated with photography very near the bed. At these low impact speed conditions, the three-dimensional nature of the real beds may lead to more efficient capture of low velocity grains, leading to a reduction in the probability of rebound.

For reasons of computational efficiency, the simulated beds used in the impact experiments were only 20 grain diameters wide; they tended therefore to possess microtopography of the order of one grain diameter. This contrasts with Bagnold's [1] observation that saltation impacts create craters with rim heights on the order of two grain diameters. In addition, preliminary measurements of microtopographic profiles with lengths of many thousand grain diameters on actual sand beds confirm that, even when artificially smoothed, such surfaces have micro-roughness at wavelengths much greater than $10D$ with amplitudes of up to $4D$. Impacts with the upwind sides of these bumps should reduce the number of rebounds to be expected, and can even result in occasional backwards-directed ejecta (or rebounds), Fig. 8, as observed in physical experiments (B. B. Willetts, personal communication, 1988). In many cases, impacts of grains with the granular surface result in rebound angles that appear much as if the surface was rigid (Werner and Haff [16]), the angle of rebound being determined largely by the surface geometry, specifically the orientation of the surface normal. A knowledge of the microtopographic profiles of the real surfaces, such as illustrated in Fig. 2, was used to calculate such surface normals for a range of possible impacts with known grain size and impact angle. The resulting reflections from the surface show distributions similar to those resulting from our calculated rebounds, and show the expected dependence of the distributions on impact angle and impacting particle diameter. Although our calculations are restricted to single grain size beds and to similarly sized impacting grains, these profiles allow estimation of the effects of variable impacting particle diameters. Larger grains impacting a finer surface will perceive a smoother surface, and should result in a lower angle rebound, while finer grains encountering a coarser surface ought to rebound at greater angles. The geometrical argument presented here will be altered by the fact that larger particles will create larger disturbances of the surface; simulations of these events are needed to clarify the relative roles of the geometrical and dynamical effects.

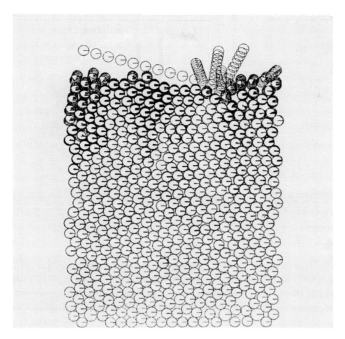

Fig. 8. Simulation of impact of $D = 0.32$ mm grain (coming from left to right) into bed of 500 similar grains, showing geometry of bed impact site is such as to cause backwards scattering of the incident particle. Several other grains are ejected at significant speeds, and the bed is influenced to several grain depths, as evidenced by fuzzing of grain boundaries in the 21 overlapped frames depicted. Note the wrap-around boundary conditions are evident in the disturbance of grains near the left side of the view; these are actually quite close to the impact site

The statistics of the rebounding and splashed grains are effectively mingled for low impact velocity events, but become more distinguishable at higher velocities. In all cases, however, an event may be most easily characterized by the sum of these two separate populations, one representing the rebound, the other the splashed grains. We follow this approach in condensing the results of simulations into a useful analytical representation of the grain-impact process to be used in the saltation simulations to follow.

8 Splash function

The results of the single impact simulations are condensed into fitted analytical expressions for the number of ejecta and the probability distribution of their ejection velocities, as functions of impact velocity (Anderson and Haff [4]). As we found little change in the properties of the grain splash over the relevant range of expected angles of impact, we have ignored any dependence on impact angle. The number of rebounding particles from a single impact of velocity V_{imi} in each of j ejection velocity "bins", labelled V_{0j}, can be fit by a Gaussian distribution

$$N_r(V_{0j}) = P_r \exp \left[-(V_{0j} - bV_{\text{imi}})/cV_{\text{imi}}^2 \right] dV_0 \tag{9}$$

where $a = 0.95$, $b = 0.56$ and $c = 0.2$, and the probability of rebound, P_r, is as previously defined above (see Fig. 9). The distribution of the number of splashed grains as a function of ejection speed, on the other hand, can be fit by an exponential curve, Fig. 9, this time

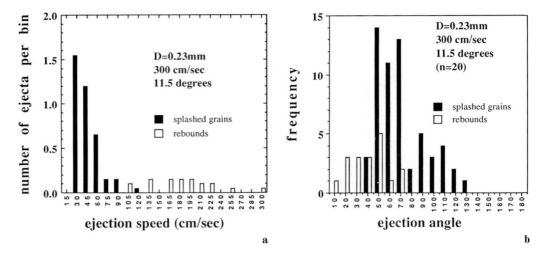

Fig. 9. a Cross section through splash function for a specific impact speed, 300 cm/sec, for 11.5 degree impact of $D = 0.23$ mm grains into a bed of similar grains. Number of grains ejected in each of several ejection speed "bins", of width 15 cm/sec. Distribution of rebounds (white bars) can well be fit by a Gaussian distribution with mean $\sim 55\%$ of the impact speed, and standard deviation $\sim 20\%$ of impact speed; sum over all rebound bars is $\leqq 1$, meaning a maximum of one grain can rebound per impact event. Distribution of splashed grains (solid bars) may be fit by an exponential function of mean and standard deviation $\sim 15\%$ of impact speed in this particular case; sum over all splash ejecta reveals that on average 3.8 grains are ejected. **b** Frequency distribution of rebound and splash angles for same case as illustrated in a), with a total of 20 impacts represented. The mean rebound angle of $38°$ and mean splash angle of $70°$ are well fit with Gaussian distributions of standard deviation $\sim 10°$

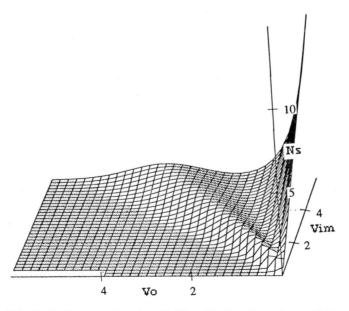

Fig. 10. Full splash function $N_s(V_{im}, V_0)$ based on the statistics of single grain size impacts from Figs. 3-9. Ejection speed bin size 20 cm/sec. V_{im} and V_0 axes in m/s. Elongated ridge running diagonally represents rebounds; curtain of increasing height (and therefore integral) represents splashed grains. Region in which V_0 is greater than V_{im} is energetically illegal; ejection speeds can never exceed impact speeds

scaled by the expected number of grains ejected per impact, $[fV_{\mathrm{imi}}]^{\gamma}$

$$N_e(V_{0j}) = [fV_{\mathrm{imi}}]^{\gamma} \exp\left[-V_{0j}/h(V_{\mathrm{imi}})^k\right] dV_0 \qquad (10)$$

where $f = 1.75$, $\gamma = 1.0$, $h = 0.25$, and $k = 0.3$. In each case dV_0 is the width of the ejection velocity bin, 0.1 m/s. The total number of grains leaving the bed in each velocity bin subsequent to a single impact of velocity V_{imi}, then, is the sum of these two expressions: $N_s = N_r + N_e$. The full splash function $N_s(V_{\mathrm{im}}, V_0)$ depicts not only the dependence of the number of ejecta on the ejection velocity, but on the impact velocity as well, Fig. 10. The sum over the all V_0 bins for a particular V_{im} represents the mean total number of grains leaving the bed subsequent to an impact of speed V_{im}, including both the rebound and the splashed grains.

Ejection angles are chosen in accord with the correlation between launch angle and launch speed (see Figs. 3, 4, 6, 7). Low V_0 grains (30—40 cm/sec; typical ejected grains) are assumed to be launched at 70—80°, while high V_0 grains (characteristic of rebounds) were launched at 35—45°. The model results are not sensitive to the choice of functional dependence of the launch angle on launch speed.

9 Modification of the wind velocity profile

As grains are accelerated by the force of the wind, they impose an equal and opposite force on the wind. The force per unit volume on the wind was calculated as a function of height and used to alter the effective stress available to shear the air at all levels in the flow. Prediction of particle concentration, mass flux, and kinetic energy flux caused by wind requires accurate calculation of particle trajectories, which in turn requires a detailed knowledge of the wind velocity profile. A theoretical model for the magnitude and vertical distribution of the drag imposed upon the wind by saltating grains is necessary to establish the relative magnitude of the three contributions to the overall drag over a mobile sand bed: saltating grains, stationary grains on the bed, and form drag due to ripples. In the eolian sediment transport system, the vertical region within which the velocity profile is modified by sediment transport is of the order of centimeters, which is great enough to allow detailed measurement of the velocity structure.

In the lowest 10 m of the atmospheric boundary layer the shear stress, τ_b, is approximately constant. Within this region, but well above the heights of the roughness elements, a logarithmic velocity profile is expected to exist, dependent upon a single velocity scale, u_*, and a single length scale, z_0. The velocity scale is the shear velocity, u_*, defined as $(\tau_b/\varrho_a)^{1/2}$, where ϱ_a is the air density. The shear stress, in turn, is governed by larger scale atmospheric circulation, driven by pressure gradients imposed by differential solar heating of the atmosphere. In the absence of large bedforms, the length scale setting the logarithmic velocity profile is proportional to the diameter, D, of the grains in the bed: $z_0 = D/30$ for uniform grains.

Sediment transport alters the effective roughness of the bed in two ways: first, the horizontal acceleration of transported grains extracts momentum from the wind, and second, the formation of ripples in the bed imposes a form drag on the wind. The resulting sediment transport roughness, z_{ost}, measured by extrapolating the velocity profile outside the sediment transport region to the $U = 0$ axis, has been shown to depend on the shear velocity (after Owen [11]; see also Rasmussen and Mikkelsen [20])

$$z_{\text{ost}} = \alpha \, \frac{u_*{}^2}{2g} \tag{11}$$

with $\alpha \sim .02$, where $\frac{u_*{}^2}{2g}$ represents, in the absence of drag, the height to which a grain with initial vertical velocity u_* is propelled.

Wind velocity measurements within the saltation region show significant deviations from the logarithmic profile. Velocity gradients nearest the bed are reduced from those expected by extrapolating the outer flow toward the bed, and result in plots of wind profiles against ln z that are convex upward. This curvature led Bagnold [1] to define a "kink" in the wind profile, which occurs typically on the order of one to a few centimeters above the bed. As stressed by Gerety [21], such near-bed deviations from the logarithmic velocity profile during sediment transport (due either to the sediment transport itself, or to the immaturity of the boundary layer at the bed of the tunnel) have led to incorrect assessment of the shear velocity during experimental work in wind tunnels. All too often, the slope of the least-squares fit to the $u - \ln(z)$ plot, from which the shear velocity is calculated, incorporates at least in part the region in which saltating grains are expected to impose a systematic departure from the simple logarithmic profile. Correct assessment of the shear velocity is essential, for instance, in the development of the correct functional dependence of the total mass flux on the shear velocity (e.g., Bagnold [1]; White [22] and compiled in Greely and Iversen [23]).

A full model of the modification of the wind profile by saltating grains must therefore be able to predict both this curvature of the profile, and the altered effective roughness of the bed. As emphasized by Ungar and Haff [18], a complete steady state saltation model must include an iteration procedure in which an altered wind profile will change the suite of particle trajectories, which in turn result in a new wind profile, etc.

Although it has long been recognized that the modification of the wind profile results from the extraction of momentum from the wind by the saltating grains, previous attempts to incorporate this effect have suffered from oversimplification of the saltation process. Thus, Owen [11] assumed that all particles trace identical trajectories. It is now clear, however, that the probabilistic nature of the grain-bed interaction leads to a broad distribution of initial conditions for particle trajectories, and to large gradients in particle concentration, mass flux, and kinetic energy flux with height above the bed (e.g., Anderson and Hallet [5]; Anderson [24], [25]; Jensen and Sørensen [26]; Sørensen [27]; Willetts and Rice [7], [8]). This structure is expected to be reflected in the profile of the force imposed on the wind by the horizontal acceleration of transported grains.

Two recent attempts have been made to calculate wind profiles during sediment transport (Ungar and Haff [18]; Sørensen [27]). Ungar and Haff confine their calculations to the simplest possible case that retains all the important elements of the saltation problem. At any shear velocity, they force their solution to yield only one particle trajectory in steady-state. As the grain-bed interaction, characterized by the "splash function", is independent of wind speed, i.e. the same lift-off velocity is retained for a particular grain impact velocity no matter what the wind structure is, the single trajectory allowed at each shear velocity must have the same impact velocity. This requires that each such trajectory experiences the same net acceleration by the wind. They argue, therefore, that in accord with Bagnold [1] "no matter how hard the wind is made to blow ... the wind velocity at a height of about 3 cm remains almost the same. Moreover, at levels still closer to the ground the wind velocity actually falls as the wind above is made stronger." Ungar and Haff's computed profiles show this behavior. It remains to be seen whether similar results can be obtained for a more realistic range of particle trajectories and a more detailed treatment of the grain-bed interaction, or "splash function".

A range of particle trajectories was introduced by Sørensen [27] in his treatment of wind velocities during sediment transport. The principal contrast between his formalism and that of the present paper lies in the nature of the postulated "closure" relation between the fluid stress and the shear rate, discussed further below.

10 Momentum equation for the air

We seek an expression for the momentum of the air during saltation that both takes account of the momentum extracted by the grains during saltation, and reduces properly to the "law of the wall" appropriate for the planetary boundary layer in the absence of saltation. Following the work of Ungar and Haff [18], we first formalize the effect of saltating grains on the air, which results in the establishment of an extra body force term in the Navier-Stokes equation for the force balance on a parcel of air. We then seek an appropriate form for the distribution of this body force with height above the bed. Finally, we suggest a possible closure relating the turbulent stresses to the local mean wind shear that remains consistent with the law of the wall in the absence of saltation.

Within the saltating curtain, the assumption of constant shear stress used to construct the law of the wall breaks down, as the acceleration of massive grains imposes an additional force on the wind. Following Ungar and Haff, we identify a horizontal body force on the wind, $F_x(z)$, acting in the upwind (negative-x) direction due to the acceleration of the grains by the air. This appears explicitly in the turbulent Navier-Stokes equations as an additional force term operating in the negative-x direction

$$\varrho_a \frac{\partial \boldsymbol{u}}{\partial t} + \varrho_a \boldsymbol{u} \cdot \nabla \boldsymbol{u} = -\nabla p + \nabla \cdot \tau_t - \varrho_a \boldsymbol{g} - \boldsymbol{F}_x \qquad (12)$$

where ϱ_a is the air density, u is the mean horizontal wind velocity, g is the acceleration due to gravity, and τ_t is the turbulent (Reynolds) shear stress, representing the flux of momentum via correlations in the velocity fluctuations. Given staedy ($\partial/\partial t \sim 0$), horizontally uniform flow ($\boldsymbol{u} \cdot \nabla \boldsymbol{u} \approx 0$), and making boundary layer approximations ($\partial/\partial z \gg \partial/\partial x$, $\partial/\partial y$), the equation for momentum in the downwind (positive-x) direction collapses to

$$\frac{\partial \tau_t(z)}{\partial z} = F_x(z) \qquad (13)$$

In the absence of saltating grains the right hand side vanishes, and the first integration yields $\tau_t = $ constant. After identifying this constant as the shear stress imposed by the exterior flow, $\tau_b = \varrho_a u_*^2$, making the common closure hypothesis that the turbulent stresses may be identified as the product of an eddy diffusivity, K, with the strain rate, $\partial u/\partial z$, and making the further assumption that the eddy diffusivity varies linearly with height, $K = k u_* z$, where k is von Karman's constant ($= 0.40$), a second integration yields the well known law of the wall, or logarithmic profile: $u = (u_*/k) \ln (z/z_0)$, where z_0 is the effective roughness of the bed. Such conditions should apply throughout the profile in the absence of sediment transport, and above the region within which grains are being appreciably accelerated by the wind during sediment transport, the difference between them being in the value of the effective roughness, z_0.

Within the saltation region, however, the stress on the wind must vary in the vertical direction as the force on the wind due to the extraction of momentum by saltating grains varies. Assuming a constant total stress available for transporting momentum of either grains or fluid across any level z, the stresses may be partitioned according to

$$\tau_b = \tau_t(z) + \int\limits_z^{z_{\max}} F_x(z)\,dz \tag{14}$$

where $z_{\max}$ is the maximum height to which a saltating grain travels in the given transport conditions.

The first term on the right hand side represents the stress available to shear the air at the level z, or the flux of fluid momentum across that level; the second, also called the "grain stress", τ_g (Sørensen [27]), represents the change in horizontal momentum of grains between their upward and downward crossings of the level z, or the net flux of horizontal grain momentum across that level. We see that as the bed is approached from above, the grain stress increases, decreasing the traction available to shear the fluid. This equation may be rewritten as

$$\tau_t(z) = \int\limits_0^z F_x(z)\,dz + \tau_{sf} \tag{15}$$

where τ_{sf} is the fluid stress at the bed, or the skin friction, part of which may be taken up by form drag due to ripples. We see then that the skin friction, or the shear stress exerted on the bed by the wind, is simply the far-field shear stress, τ_b, minus the total change in horizontal momentum of all grains ejected from a unit area of bed in a unit time,

$$\tau_{sf} = \tau_b - \int\limits_0^{z_{\max}} F_x(z)\,dz \tag{16}$$

For a given exterior wind condition, characterized by τ_b, as the body force diminishes in magnitude, more fluid stress is made available at the bed for aerodynamic initiation of saltation; and conversely, as body force increases, the shear at the bed decreases. Owen [11] hypothesized that the self-regulatory nature of eolian saltation arose from such a feedback, with skin friction held near the threshold shear stress necessary to maintain particles in transport by impacts, Bagnold's impact threshold. It remains to test this hypothesis quantitatively.

We now seek an equation for the velocity gradient as a function of height, which when integrated will yield a velocity profile in equilibrium with sediment transport. Given the above relations between far field shear stress, skin friction, and grain stress, we require both the body force profile, $F_x(z)$, and a constitutive relation between the turbulent stresses and the mean velocity gradient. For simplicity, we again postulate an eddy diffusivity closure: $\tau_t = \varrho_a K \, \partial u / \partial z$, and retain the linear dependence of the eddy viscosity with height, $K = k u_* z$, that gives rise to the logarithmic velocity profile in the absence of sediment transport. However, referencing, u_* to the total stress, τ_b, is no longer appropriate. We define a local, or effective shear velocity, referenced to the local turbulent stress, τ_t, as

$u_{*\text{eff}}(z) = (\tau_t/\varrho_a)^{1/2}$. Then, from $\tau_t = \varrho_a k z u_{*\text{eff}} \, \partial u/\partial z$, we have

$$\frac{\partial u}{\partial z} = \frac{1}{kz} \left[\frac{\tau_b - \int\limits_{z}^{z_{\max}} F_x(z) \, dz}{\varrho_a} \right]^{1/2} \qquad (17)$$

Note that above the region within which grains accelerate, i.e., when $z > z_{\max}$, or in the absence of sediment transport altogether, i.e. when $F_x = 0$ for all z, the numerator becomes $(\tau_b)^{1/2}$, and the rate of shear becomes $\partial u/\partial z = u_*/kz$, which again yields the logarithmic profile, as required. Given the form of the force profile at any time in the evolution of the saltation population, derived in the next section, this equation may be numerically integrated to yield the corresponding wind velocity profile.

Note that irrespective of the actual form of the force profile, the fluid stress, τ_t, and hence the effective shear velocity, will increase monotonically with height, giving rise to convex upward $u - \ln(z)$ plots. Those profiles showing inflections in semi-log space, as some of Bagnold's [1] early profiles do, are suspect; the stress available to shear the fluid, and hence the effective shear velocity, should increase monotonically away from the bed, yielding convex-upward wind velocity profiles. Inflections resulting from an initial decrease in effective shear velocity, followed by an increase in shear velocity, should not exist.

11 Force on the wind due to saltating particles

Since the force on the wind is directly related to the drag force experienced by the saltating particles (Newton's third law), we compute the horizontal component of the force on the wind, $F_x(z)$, by looking at the drag on individual grains as they move along their trajectories (Anderson and Hallet [5]; Anderson [25], Sørensen [27]). Figure 11 illustrates the horizontal component of the force of the the wind on a particle as a function of position along the trajectory, $f_x(z) = Ma_x(z)$, where a_x is the instantaneous horizontal acceleration of

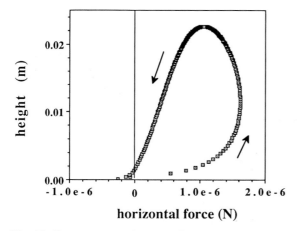

Fig. 11. Instantaneous horizontal force on a grain as a function of position along its trajectory. Upward arrow denotes ascending portion of trajectory; downward arrow denotes descent. Highest force on the particle (and therefore on the wind) occurs in ascent. Slightly negative force just before impact reflects that the particle here has more horizontal speed than the wind, and is therefore slowed by it

the grain and M is the particle mass. The highest instantaneous force is attained early in the ascending limb of the trajectory, before the particle has been appreciably accelerated by the wind, and where the relative velocity between the particle and the air is therefore the greatest. The force on the particle becomes negative shortly before impacting the bed, indicating that the particle there is travelling faster than the wind.

Summing over the ascending and descending limbs of the trajectory, and assuming a single particle is ejected per unit area of bed per unit time, i.e. $N_1 = 1$, with initial vertical velocity w_0, results in the "identical trajectory" force per unit volume on the wind

$$G_x(z|w_0, N_1) = N_1 M \left[\left(\frac{a_x}{|\langle w(w_0) \rangle|} \right)_+ + \left(\frac{a_x}{|\langle w(w_0) \rangle|} \right)_- \right] \tag{18}$$

where $\langle w \rangle$ is the mean vertical particle velocity as it crosses the height element $(z - dz, z)$, and the $+$ and $-$ denote upward and downward limbs of the trajectory, respectively (Ungar and Haff [18]; Anderson [25]). The upwind direction of the force is left implicit. The "identical trajectory" force profile, displays a distinct maximum at the top of the trajectory, similar to the maxima in particle concentration, mass flux, and kinetic energy flux profiles reported earlier (Anderson and Hallet [5]; Anderson [24]). Although the instantaneous horizontal force on the particle peaks approximately one third of the way up the ascending limb of the trajectory, Fig. 11, most of the particle's horizontal acceleration occurs near the top of the hop, where its vertical velocity is very low, and therefore where it spends the most time. This is the origin of the velocity factors in the denominators in the above equation. A similar argument explains the maximum at the top the particle path in the concentration, mass flux, and kinetic energy flux profiles for "identical trajectory" models (Anderson and Hallet [5]; Anderson [24]).

The probability density of the vertical liftoff velocity, $p(w_0)\,dw_0$, and the actual number of particles ejected per unit area of bed per unit time, N, are now introduced to yield an integral equation for the total horizontal body force per unit volume on the wind due to the presence of saltating particles

$$F_x(z) = \frac{N}{N_1} \int_0^\infty p(w_0)\, G_x(z|w_0, N_1)\, dw_0. \tag{19}$$

This is illustrated in Fig. 12, for a particular choice of grain size and u_*. In earlier work (Anderson [25]), force profiles and the resulting wind velocity profiles were calculated using reported total mass fluxes and assumed forms for the probability distribution, $p(w_0)$, to constrain the values of the total ejection rate, N, in the above equation. These calculations revealed that the effect of the form of the probability distribution of liftoff velocities dominates over the shape of the "identical trajectory" force profile in determining the result of the integration. Force profiles decline sharply above the bed, with a scale height that is on the order of 1—3 cm. The effect of transported grains on the wind profile, for moderate winds, should essentially vanish by 5 cm above the bed, as is observed. As this scale height is a reflection largely of the probability distribution of initial velocities, one expects this will be strongly influenced by grain size, and by the external forcing characterized by u_*.

The choice of a linearly increasing eddy diffusivity within the saltation region follows more closely the treatments of Ungar and Haff [18] and of Sørensen [27], is identical to the mixing length algorithm used by Werner [13], and diverges from that of Owen [11], who argued that a constant eddy diffusivity would be appropriate. Owen made the plausible argument that the intensity of turbulence and the mixing lengths of the turbulence

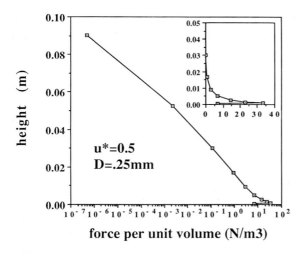

Fig. 12. Log-linear plot of the force per unit volume on the wind due to the acceleration of sand grains, at steady state under the conditions of $D = 0.25$ mm, and $u_* = 0.5$ m/s. Inset, a linear-linear plot of the same case. The force profile decays monotonically, and roughly exponentially, above a level very close to the bed, corresponding to the position (see Fig. 11) below which impacting grains are moving at speeds greater than the wind

ought both to be dominated by wakes cast by saltating particles, and should therefore be roughly constant within the saltation curtain. However, the concentration of saltating particles is on the order of 10^{-2} to 10^{-4} near the bed (Anderson and Hallet [5]; Gerety [21]; Sørensen [27]); it may therefore be expected that the nearby presence of a continuous rough bed will dominate over the wakes cast by these sparsely distributed particles in setting the length scale of the turbulence. The use of a linearly varying eddy diffusivity also allows a much simpler feathering of the saltation region with the constant stress region above, rather than requiring an abrupt change in the nature of the turbulence. This explicitly recognizes that the saltation region is neither physically isolated, nor even easily identified in the real world. The character of the saltation curtain is such that all profile quantities, including the body force on the wind imposed by the accelerating particles, fall off rapidly away from the bed, as a consequence primarily of the heavily skewed nature of the probability density of liftoff conditions resulting from the complex grain-bed interaction. It is a sediment transport boundary layer.

That the shear velocity be tied to the *local* fluid stress, τ_t, rather than the total or far-field stress, τ_b, makes the present treatment different from those of both Ungar and Haff [18], and Sørensen [27]. Such an hypotheses has proven fruitful in the analysis of aqueous systems, where the form drag due to multiple sets of bedforms is modelled with a corresponding number of matched logarithmic profiles, each characterized by a shear velocity referenced to the spatially averaged form drag associated with that particular bedform scale (e.g., Smith and McLean [28]). The present expression for the wind shear identifies both the magnitude of the stress due to saltating grains, and its profile, allowing a smoothly varying wind profile rather than a matched logarithmic profile.

A further difficulty with Owen's [11] formulation of the wind profile within the saltation region is that the no-slip condition is left unsatisfied. Using his formula, wind velocities at the bed remain of the order $7-8u_*$. Although this has prompted earlier workers to justify use of a constant wind velocity ($= 8.5u_*$) in the saltation layer, thereby simplifying trajectory calculations, the present formulation is more realistic, is computationally manageable, and provides a profile of most physical quantities of interest in sediment transport mechanics.

12 Saltation model

The actual particle motion through the air is handled as follows. Because of computational limitation, particles are allowed to leave the bed with one of only ten distinct velocities; i.e., the distribution of initial conditions is discretized. These are chosen such that the lowest liftoff velocity trajectory reaches a height δ equal to approximately one grain diameter, while that with the highest liftoff velocity reaches a height above which little saltation flux is expected. This latter limit was chosen to be $w_0 = 5u_*$, meaning that typical trajectory heights for these grains is $\delta = 25u_*^2/2g$, or roughly 1.2 m for $u_* = 1.0$ m/s. Little grain flux is expected above these heights during saltation of sand in air. The intervening liftoff velocity "bins" are distributed logarithmically in order to maximize the profile information in the lower portion of the saltation curtain, where most of the grains travel. At any instant in time, each liftoff velocity bin $(V_0)_i$, contains N_i particles.

13 Collision list

Participation in the splash process requires impact with the surface, which occurs with decreasing frequency as the initial velocity of the trajectory increases. (As noted previously, trajectory times are well estimated by $\tau_{\text{hop}} = \psi(2w_0/g)$, where w_0 is the vertical component of the liftoff velocity, and ψ is a coefficient of value less than but close to 1, only weakly dependent upon the initial velocity of the grain.) To account for variability in trajectory durations, a "collision list" is established. For each trajectory class a list is maintained of the number of grains expected to be impacting the surface at each time increment in the future, their launch times and hop durations being known. At each time step, then, the collision list for each trajectory class is interrogated to reveal the number of impacts in each trajectory bin. Because the wind velocity profile may be changing during a long trajectory, the impact velocities are assessed by calculating the trajectories using the wind velocity profile averaged over the hop time of the trajectory. The number of splashed particles in each of the liftoff velocity bins is then calculated from the splash function. Finally, the collision list is updated in accord with this new set of ejecta, and the entire collision list is moved forward in time one step to reveal the new numbers of impacts in each bin to be expected upon the next time step.

Other than this accounting for variable trajectory times via the collision list, the calculation runs according to the box diagram presented in earlier work (Anderson and Haff [4]). Time is discretized at two levels. The time step associated with the reassessments of the wind profile and the grain splash is determined by the shortest trajectory time, or roughly $4(D/g)^{1/2}$ (the hop time of a grain reaching a height of $2D$). A second, much shorter time step is needed to assure that each trajectory is calculated with sufficient precision to account correctly for the impact velocity and the various profile quantities of concern. Were we to have to calculate the actual grain dynamics of the bed upon each and every impact with the bed, a third, much shorter time step would be involved, reflecting the elastic material properties of the bed grains. The use of the splash function to represent the response of the bed to impacts removes this necessity.

14 Full simulations

Inputs required for the full simulation of the eolian saltation system are grain diameter and density, an initial wind velocity profile, the constants determining the splash function, and the coefficient of proportionality between the excess shear stress and the aerodynamic entrainment rate, ζ. The initial wind velocity profile is taken to be logarithmic. For a flat bed, the roughness height will be proportional to the diameter of the sand grains in the bed. In his experiments on the effect of sand movement on the surface wind, Bagnold [1] produced wind profiles over a wetted sand bed that was previously "not only pitted with tiny bombardment craters a few grain diameters in size, but was made to undulate in the usual flat transverse ripples". The resulting $u - \ln(z)$ profiles for $u_* = 0.20 - 0.62$ m/s show little or no curvature between 2 mm and 10 cm above the bed, and all yield roughness heights closely approximated by $z_0 = 2D/30$, where $2D$ is roughly the mean height of the impact craters (Bagnold [1]). Accordingly, the initial roughness height was chosen to be $2D/30$. (We note that this is also roughly in accord with recent findings of Whiting and Dietrich [29] on natural mixed grain size fluvial bed surfaces, where $z_0 = D/10$.) The shear velocity, u_*, in turn sets the initial shear stress at the bed, $(\tau_a)_0 = \tau_b = \varrho_a u_*^2$, to be used in subsequent recalculations of the stress and wind profiles.

At each time step, the collision list is interrogated to reveal what number of grains in each of the 10 trajectory classes is due to impact the bed. The actual trajectories of these grains are then calculated using the wind profile averaged over the hop time (approximated by $\tau = 2w_0/g$, where $w_0 = V_0 \sin\theta$). The impact speed of each trajectory is then used as input to the splash function, which returns the number of ejecta per impact (including rebounds) in each ejection speed class. The wind velocity profile is then adjusted in accord with the calculated body force profile, $F_x(z)$. The number of grains leaving the bed at this instant due to aerodynamic forces is assessed, and added to those leaving the bed in the lowest speed class due to impacts. The collision list is updated correspondingly. Other profile quantities, such as mass flux and concentration are recorded, and the clock is stepped forward. Simulations are run until a steady state is achieved, a state characterized by little or no change in the mass flux, in the wind velocity profile, or in the collision list from one time step to the next.

Examples of the simulated evolution of the saltation population and of associated quantities of interest are shown in Figs. 13—15. Initial entrainment is entirely aerodynamic. Entrained grains gain horizontal momentum from the wind, and impact the surface with velocities such that a small proportion rebound with greater initial velocities than they had at entrainment; few subsequent grains are splashed at this stage. An initial delta function probability distribution of liftoff velocities therefore evolves to a broader probability distribution, filling out into the higher velocities. The grains with higher liftoff velocities are airborne for longer periods of time before impacting and contribute more strongly to the splash ejection of other grains. The full range of possible ejection velocities is populated only after many tens of short-trajectory times. At this point the total number of grains in transport begins to grow rapidly, the highest impact velocity grains being most efficient in splashing grains into the airstream. The resulting roughly exponential growth is curtailed only when the extraction of momentum from the wind is sufficient to alter significantly the wind velocity profile, which in turn alters the impact velocities of the longer grain trajectories, Fig. 13.

The system eventually reaches a steady state characterized by a specific total mass flux, Fig. 14, an equal number of impacting and ejected grains, Fig. 13, and a stationary wind velocity profile, Fig. 15. The overshoot of the steady state appears to be due to the time lag associated with the 0.2 to 0.3 second hop times of the most energetic trajectories,

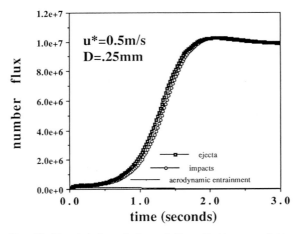

Fig. 13. Simulated evolution of the saltating population, showing the number of grains leaving the bed per square meter per second due to both impacts and aerodynamic entrainment, and the number of grains impacting the bed. At steady state the ejection rate equals the impact rate (at on the order of $10^7/\text{m}^2-\text{s}$), and aerodynamic entrainment rate has fallen to zero. Steady state is achieved in roughly 2 seconds

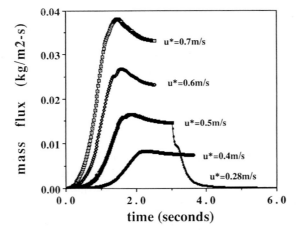

Fig. 14. Simulated evolution of mass flux for each of 4 cases with u_*'s of 0.4—0.7 m/s. In all cases $D = 0.25$ mm, and the grain density is that of quartz, $\varrho = 2\,650$ kg/m^3. In each case steady flux is achieved in order 1—2 seconds, slightly lower times for higher shear velocities. Note also the single simulation demonstrating hysteresis in the system: after steady state is achieved at $u_* = 0.5$ m/s, the driving stress is reduced to lower than that necessary to entrain grains aerodynamically (i.e., $u_* < u_{*c}$, which for this case if taken to be ~ 30 cm/sec). A new very low but finite steady state flux is achieved in ~ 1 sec

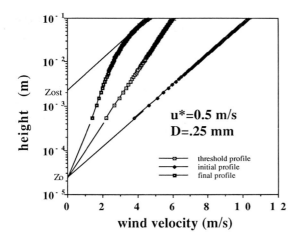

Fig. 15. Wind velocity profiles on log-linear plot shown at initial condition and final condition (steady state). Also shown is the threshold wind velocity profile below which no aerodynamic initiation of trajectories is expected. Note final profile has a near surface gradient that is less than the threshold, meaning the fluid stresses are insufficient to entrain grains from the bed. Straight line extrapolations of wind velocity from above the saltation layer to $u = 0$ demonstrate that the $u = 0$ intercept increases from the initial to the steady state, from z_0 to z_{0st}, representing the increased surface roughness perceived by the flow

which are responsible for the majority of the ejections from the bed. The steady state mass flux is within the range of fluxes measured in wind tunnel studies for the same combination of shear velocity and grain size. Whereas the initial saltation population is entirely aerodynamic, the steady state saltation population for most imposed shear velocities contains no aerodynamically entrained grains. As the population of splashed and rebounding grains increases, the fluid shear stress at the bed is reduced, with a corresponding decrease in the rate of aerodynamic entrainment. The steady state shear stress at the bed ultimately decreases to slightly below the critical shear stress for aerodynamic entrainment. The effective roughness of the bed is simultaneously greatly increased, as reflected in the rise of the $U = 0$ intercept (from z_0 to z_{ost}), Fig. 15. This change in bed roughness is in accord with numerous measured wind velocity profiles during saltation experiments, as summarized by Owen [11]. The resulting steady state profile of mass flux, Fig. 16, displays the characteristic rapid decrease above the bed, implying that the system has envolved to produce a realistic probability distribution for the initial trajectory velocities.

By varying the initial shear velocity, several such simulations were performed (Anderson and Haff [4]) to produce a mass flux relation that is broadly similar to those derived empirically: above a threshold shear velocity ($u_{*c} = (\tau_c/\varrho_a)^{1/2}$), the flux increases as a power of ($u_* - u_{*c}$), Fig. 17b.

Further calculations demonstrated yet another long-recognized feature of eolian saltation: hysteresis. Once steady state had been established using a particular shear velocity, the externally imposed shear stress (and hence u_*) was diminished (in stepped changes) to below that necessary to entrain grains aerodynamically. For choices of this shear velocity down to approximately 0.75 of the aerodynamic threshold shear velocity, u_{*c}, a low but steady and finite flux was achieved, Fig. 17a. This reflects the hysteresis in the system that forced Bagnold to define two threshold velocities: a fluid threshold, and an impact threshold, the latter being on the order of 0.8 of the former.

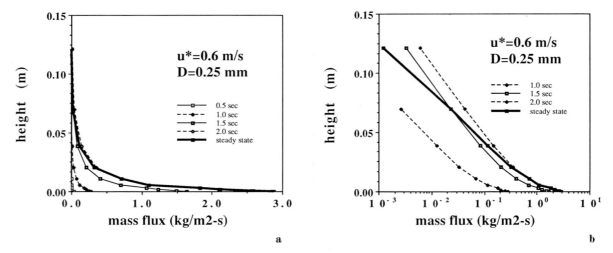

Fig. 16. a Mass flux profile captured at 0.5 second intervals en route to steady state profile (bold line) at ∼ 2.5 seconds. Note in all cases the monotonic decline in mass flux with height above the bed, reflecting a probability distribution of liftoff velocities heavily skewed toward the low velocity ejecta. **b** Same as (a) but on log-linear plot. Note that extrapolation of the exponential distribution seen well above the bed would lead to large (10-fold) underestimate of flux nearest the bed, in the lowest 1—2 cm

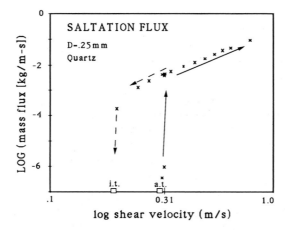

Fig. 17. Log steady state mass flux *vs.* log u_*, showing clearly the hysteresis in the system. Upon increasing shear velocities (solid arrows), the system turns on extremely rapidly at values only slightly above the aerodynamic threshold (a.t.; here 0.30 cm/sec), then increases at a lower rate, but still nonlinearly (with power ∼ 3—4). Upon decreasing u_* (dashed arrows), this steady state mass flux *vs.* u_* path is retraced until the aerodynamic threshold, below which finite mass fluxes are allowed (see, for instance, Fig. 14) until a second, lower threshold is crossed (the impact threshold, i.t., here roughly 0.27 m/s), at roughly 0.7—0.8 of the aerodynamic threshold

The response time of the saltation system appears to be on the order of 1 to 2 seconds, or several long-trajectory times (Anderson and Haff [4]; Anderson [30]). For the cases run to date, it appears that the response time is a weak function of the shear velocity, being longer for lower shear velocities. A knowledge of the response time, which is difficult to obtain from wind tunnel experiments, will allow us to treat the problem of predicting total mass fluxes in variable winds.

15 Discussion and conclusions

By introducing a realistic representation of the granular splash process, and of the wind velocity feedback process, we have been able to reproduce the essential features of eolian saltation. These include the details of profile quantities such as mass flux, concentration, and wind velocity, as well as the complex, highly nonlinear and hysteretical dependence of integrated quantities such as the total mass flux on the driving stress. Calculated steady state mass flux profiles monotonically decline in value above the bed, reflecting the evolution of the probability distributions of liftoff velocities to those dominated by low ejection velocity grains. This both confirms many earlier experimental findings, and removes the impediment encountered in earlier theoretical work in which a specific form of the probability distribution (e.g., Anderson and Hallet [5]) had had to be assumed in order to calculate these profiles. Jensen and Sørensen [26], in their estimation of the launch velocity probability distributions from measured mass flux profiles, had also to make certain assumptions about the general form of the distribution. Clearly, the dominance of the lower liftoff velocity grains reflects the details of the splash process, in which many low energy grains may be ejected by single high energy impacts, and in which the probability of rebound decreases dramatically as the impact speed declines.

The magnitudes of the integrated fluxes (which assign the scales to each of these profiles) are determined by the wind velocity feedback. This feedback limits the total number of grains either impacting or being ejected from the bed at steady state (at steady state these are equivalent). This ejection rate had in the past been estimated from measured mass fluxes in wind tunnels, but here falls out as a result of the full calculation.

While the current model therefore represents considerable progress in our understanding of the eolian saltation process, and in our ability to model it quantitatively with a few empirical parameters as possible, this progress has been made at the expense of several limiting assumptions. Several are related to the nature of the granular splash. We have not addressed the possibility that the granular splash process may depend upon the "state" of the bed, i.e., on how well or poorly packed it is. It has been suggested (Sørensen [31]) that the fluffing of the bed under intense saltation may provide an important feedback on the saltation population. Our calculations of granular splashes were performed under conditions of a freshly settled bed that had not as a whole been micro-tamped by previous saltation impacts, nor in which the surface was allowed to be fluffed by saltation. In addition, the beds used in the simulations are two-dimensional; the nature of the packing in real beds may make the bed effectively softer. The two-dimensionality of the grain splash calculations at present also precludes attention to the lateral component of the granular splash process, which is important in the dispersion of grains in saltation, and in the establishment of laterally extensive ripples. Nonetheless, the quantitative correspondence of the granular splashes presented here with those observed in the physical world is encouraging.

We currently assume the wind has reached a steady state throughout its profile, meaning that no longer is momentum being fed into the deficit region associated with the saltation curtain. The implicit spatial and temporal dependence associated with the transient effective change in the roughness of the bed, which operates with time scales long compared to the variability of the wind, is therefore ignored. While this may affect the response times and the total mass fluxes at steady state, we expect its effect to be minor. Nonetheless, the implementation of a more realistic wind is crucial to the approach toward realism. This must be accomplished through incorporation not only of the growing boundary layer problem mentioned above, but of the gustiness of real winds as well. These alterations will have several ramifications. More aerodynamic entrainment is expected in the real world, as gusts periodically interrupt the growth of the saltation population. A length scale for the development of the saltation layer should also emerge from a fuller incorporation of the growing boundary layer. We expect that this length scale could be quite long (at least several long (energetic) hop lengths), which raises the important question of the necessary lengths of wind tunnels used in the study of eolian saltation.

It is likely, however, that the problems associated with a more realistic wind are minor in comparison with those associated with the introduction of multiple grain sizes (even two) in the saltation process. While the effects of multiple grain sizes are essentially trivial to incorporate in the algorithms for the trajectory and wind feedback processes, they add another dimension to the already complex splash function, in that the granular splash has to be assessed for any combination of impactor, whose size is discretized, and bed mixture, which runs a continuum from all-fine to all-coarse grains. Nonetheless, an advance to the treatment of at least two grain sizes is warranted, as important geological problems of dust ejection from particle mixtures, grain sorting and associated bed structures, and the ubiquitous horizontal sorting of coarse grains to the crests of ripples require treatment of the multiple grain size saltation problem.

Acknowledgements

This work was sponsored by the U.S. Army Research Office under grant number DAAL03-89-K-0089, and by the National Science Foundation (EAR-89-16102 [RSA] and EAR-89-15983 [PKH]). We greatly appreciate reviews of the manuscript by M. Sørensen and M. R. Raupach.

References

[1] Bagnold, R. A.: The physics of blown sand and desert dunes. London: Methuen 1941.

[2] Chepil, W. S.: Dynamics of wind erosion. I. Nature of movement of soil by wind. Soil Sci. **60**, 305—320 (1945).

[3] Anderson, R. S.: A theoretical model for aeolian impact ripples. Sedimentology **34**, 943—956 (1987).

[4] Anderson, R. S., Haff, P. K.: Simulation of eolian saltation. Science **241**, 820—823 (1988).

[5] Anderson, R. S., Hallet, B.: Sediment transport by wind: toward a general model. Geol. Soc. Am. Bull. **97**, 523—535 (1986).

[6] White, B. R., Schulz, J.: Magnus effect in saltation. J. Fluid Mech. **81**, 497—512 (1979).

[7] Willetts, B. B., Rice, M. A.: Inter-saltation collisions. In: Barndorff-Nielsen, et al. (eds.) Proceedings of the International Workshop on the Physics of Blown Sand, vol. 1. Department of Theoretical Statistics, University of Aarhus, pp. 83—100 (1986).

[8] Willetts, B. B., Rice, M. A.: Collisions in eolian saltation. Applications of the mechanics of granular materials. In: Geophysics, Euromech 201 Conference; Interlaken, Switzerland, Oct. 13—18, 1985 (1986).

[9] Willetts, B. B., Rice, M. A.: Collisions of quartz grains with a sand bed: influence of incidence angle. Earth Surface Processes and Landforms **14**, 719—730 (1989).

[10] Willetts, B. B., Rice, M. A.: Particle dislodgement from a flat bed by wind. Earth Surface Processes and Landforms **13**, 717—728 (1988).

[11] Owen, P. R.: Saltation of uniform grains in air. J. Fluid Mech. **20**, 225—242 (1964).

[12] Werner, B. T.: A physical model of wind-blown sand transport. Ph. D. dissertation, California Institute of Technology, Pasadena, California USA. p. 441 (1987).

[13] Werner, B. T.: A steady state model of windblown sand transport. J. Geol. **98**, 1—17 (1990).

[14] Mitha, S., Tran, M. Q., Werner, B. T., Haff, P. K.: The grain-bed impact process in aeolian saltation. Acta Mech. **63**, 267–278 (1986).

[15] Rumpel, D. A.: Successive aeolian saltation: studies of idealized collisions. Sedimentology **32**, 267–275 (1985).

[16] Werner, B. T., Haff, P. K.: The impact process in eolian saltation: two dimensional studies. Sedimentology **35**, 189–196 (1988).

[17] Werner, B. T., Haff, P. K.: Dynamical simulations of granular materials using concurrent processing computers. In: Fox, G. C. (ed.) Proceedings 3rd Conference on Hypercube Concurrent Computer and Applications, Jan. 19–20, 1988, New York: ACM 1988.

[18] Ungar, J. E., Haff, P. K.: Steady state saltation in air. Sedimentology **34**, 289–299 (1987).

[19] Werner, B. T., Haff, P. K.: A simulation study of the low energy ejecta resulting from single impacts in eolian saltation. In: Arndt, R.E.A., et al. (eds.) Advancements in aerodynamics, fluid mechanics and hydraulics. New York: Am. Soc. Chem. Eng. 1986.

[20] Mikkelsen, H., Rasmussen, K. R.: Transport profiles measured with an isokinetic trap (this volume).

[21] Gerety, K.: Problems with determination of u^* from wind-velocity profiles measured in experiments with saltation. In: Proc. International Workshop on the Physics of Blown Sand, vol. 2. Dept. of Theoretical Statistics, Aarhus University (Denmark), Mem. 8, pp. 271–300 (1986).

[22] White, B. R.: Soil transport by wind on Mars. J. Geophys. Res. **84**, 4643–4651 (1979).

[23] Greeley, R., Iversen, J.: Wind as a geological process. Cambridge: Cambridge University Press 1985.

[24] Anderson, R. S.: Erosion profiles due to particles entrained by wind: application of an eolian sediment-transport model. Geol. Soc. Am. Bull. **97**, 1270–1278 (1986).

[25] Anderson, R. S.: Sediment transport by wind: saltation, suspension, erosion and ripples. Ph. D. dissertation, University of Washington, Seattle (1986).

[26] Jensen, J. L., Sørensen, M.: Estimation of some eolian saltation transport parameters: a reanalysis of Williams' data. Sedimentology **33**, 547–555 (1986).

[27] Sørensen, M.: Estimation of some eolian saltation transport parameters from transport profiles. In: Proc. International Workshop on the Physics of Blown Sand, vol. 1. Dept. of Theoretical Statistics, Aarhus University (Denmark), Mem. 8, pp. 141–190 (1986).

[28] Smith, J. D., McLean, S. R.: Spatially averaged flow over a wavy surface. J. Geophys. Res. **82**, 1735–1746 (1977).

[29] Whiting, P., Dietrich, W. E.: The roughness of alluvial surfaces; an empirical examination of the influence of size homogeneity and natural packing. Eos **70**, 1109 (abstract) (1989).

[30] Anderson, R. S.: Feedbacks and time scales in eolian saltation. Eos **69**, 1195 (abstract) (1988).

[31] Sørensen, M.: The soft bed — a proposal for a self-limiting mechanism for saltation. Program and abstracts of a workshop on dynamics of eolian sediment transport, A.S.U., Tempe, Arizona USA, March 30, 1988, 38–39 (1988).

Authors' addresses: Dr. R. S. Anderson, University of California Santa Cruz, Earth Sciences Board, Applied Sciences Building, Santa Cruz, California 95064, U.S.A., and Prof. P. K. Haff, Department of Civil and Environmental Engineering, Duke University, School of Engineering, Durham, North Carolina 27706, U.S.A.

Acta Mechanica (1991) [Suppl] 1: 53—66

Numerical model of the saltation cloud

I. K. McEwan and **B. B. Willetts**, Aberdeen, United Kingdom

Summary. A computer model of the saltation cloud is described. Experimental results from high speed films are used to characterise the grain/bed collision. The importance of momentum exchange in determining the number of ejected grains from a collision is demonstrated. The modification of the wind velocity profile is discussed and a realistic wind profile is calculated. Also the mass flux profiles calculated compare well to their expected shape. The model attains a steady state, characterised by a steady wind and a stationary grain population, after roughly 2 seconds. The response of the total mass flux to shear velocity is approximately cubic. Finally, potential uses of the model in studying ripple formation and dust emission are discussed.

1 Introduction

Some of the most striking developments in recent years toward a full theory of saltation have occurred through the use of the computer as an experimental tool. This affords experimental control which is almost impossible in conventional experiment. Dependencies may be more easily isolated than they can in wind tunnel or field work from which undesired influences can only be eliminated with great difficulty, if at all. Yet there are dangers too. An environment may be constructed within the computer which has scant relevance to nature. Great care must be taken that in applying a reductionist approach important ingredients of the problem are not lost.

Nevertheless computer models, constructed with due regard for these considerations produce better understanding of problems such as ripple formation, grain/bed interaction, and steady state saltation [1], [2]. These models pass difficult tests set by experimental data, such as the calculation of realistic wind and mass flux profiles. The total mass fluxes calculated also show a satisfactory dependence on shear velocity and Anderson and Haff's model [1] demonstrates the characteristic hysteresis between the fluid and impact thresholds. Confidence is high that these models manifest the salient features of saltation.

The objective of this paper is to describe a further development in such models and to present some of the results achieved. It goes on to indicate intended and potential uses of the model.

2 The model

The model operates within a domain which may be considered analogous to a 'wind tunnel' of infinite height with nominal base dimensions of 2.00×0.01 m. The model has periodic boundaries meaning that grains which leave at one end re-enter at the other. The initial wind profile is logarithmic with height and the initial surface roughness is set to 1/30 of a grain diameter. The model is seeded by a small number of lift-off velocity vectors recorded from film. These grains are selected randomly from the film data and take-off at random positions along the bed. The grains are sufficient to trigger the chain reaction of grain impact in motion toward a steady state. The seeding grains are tracked forward in time until they collide with the bed and rebound, ejecting other grains. Simultaneously the wind profile is being adjusted so that the new grains travel in a wind of reduced strength. This interaction between the grains and the wind continues until the number of births[1] and deaths[2] are equal and the wind is stable, characterising the steady state.

There is no aerodynamic entrainment in this model. At steady state it is assumed that direct wind entrainment is rare and that young saltations are generated by grain impacts [1]. The incorporation of fluid entrainment will be considered in the discussion.

3 The trajectory calculation

The calculation of saltation trajectories has been reliably performed in many previous studies [1], [2], [3], [4], [5], [6]. It is widely accepted that there are four forces which govern the flight of a saltation.

1. force due to gravity
2. aerodynamic drag
3. Magnus effect
4. aerodynamic lift

Following the precedent set by Hunt and Nalpanis [6] and Sørensen [4] aerodynamic lift and the Magnus effect are omitted from the calculation. The equations of motion used in the model are

$$x'' = -0.75 \frac{\varrho_a}{\varrho_p} \frac{Ur}{d} C_d \big(x' - U(y)\big) \tag{1}$$

$$y'' = -0.75 \frac{\varrho_a}{\varrho_p} \frac{Ur}{d} C_d y' - g \tag{2}$$

where ϱ_a is the density of air ($= 1.23 \text{ kg/m}^3$)

ϱ_p is the density of quartz ($= 2650 \text{ kg/m}^3$)

U_r is the grain's relative velocity

d is the grain diameter

C_d is the drag coefficient

$U(y)$ is the time-averaged wind velocity

y is the grain's y co-ordinate

x' is the grain's x velocity

y' is the grain's y velocity

x'' is the acceleration in the x direction

y'' is the acceleration in the y direction

[1] new saltation sequences

[2] termination of saltation sequences

The drag coefficient C_d is a function of Reynold's number. If it is accepted that C_d for a sand grain may be represented by the drag coefficient for an equivalent diameter sphere then a number of empirical formulae are available for its determination [7], [8], [9]. Like White and Schulz [5] and Anderson and Haff [1] we use the result of Morsi and Alexander [9].

The wind velocity, $U(y)$ is the time averaged wind velocity at height, y. Hunt and Nalpanis [6] suggested that a tentative distinction may be drawn between suspension and saltation at a grain diameter of 100 μm. The grain diameter used in this model is consistently larger than this figure. Therefore motions of grains within the model would not be greatly influenced by the turbulent fluctuations and so the use of a time averaged wind velocity is justified. It is worthwhile to note, in passing, that the grains, on leaving the surface are divided into discrete velocity classes (or bins) so that a representative trajectory is calculated for a grain in a particular class. Inclusion of a turbulent wind would force the abandonment of this procedure in favour of tracking individual grain flights. This would increase the demand on computation beyond present and foreseeable resources.

4 The splash function

The term splash, chosen to represent the action of a grain striking the bed has won its way into general terminology. It first appeared in Owen [3] and has since been incorporated into the concept of the *splash function* [10]. A saltation approaches the bed at an angle of 10—15° and ricochets, typically at an angle of 30°, with a reduced velocity. The exchange of momentum with the bed effects the ejection of a number of bed grains with velocities an order of magnitude less than that of the impinging grain. It is these ejected grains or reptations that suggest the notion of a saltation splashing through the surface.

The splash function remains poorly defined. The inputs to the function are likely to include impact angle, impact velocity, the mass of the impact grain and the bed angle [11], [12]. The outputs are the velocity vectors of the ricochet grain and the number, masses and velocity vectors of the ejected grains.

The splash function used in this model was derived from a subset of data presented by Willetts and Rice [11]. The data consisted of observations from 98 collisions of a medium grain size fraction on a bed consisting of three fractions: fine (150—250 μm), medium (250—355 μm) and coarse (355—600 μm). In most collisions the course of the impinging grain could be followed with some certainty through the bed until it re-emerged. Therefore there is little doubt that in these cases the ricochet grain is the same as the impacting grain. However it was not possible to ascertain the size of the grains ejected from the bed. It is primarily because of this experimental difficulty that the saltation model suffers from the shortcoming of dealing only with uniform grain size. Nevertheless despite this shortcoming and the smallness of the data sample some encouraging results emerge from the splash data.

5 The ricochet grain

The magnitude of the rebound velocity vector was found to have a roughly linear dependence upon the magnitude of the incident grain speed. The envelope[1] demonstrating this relation is shown in Fig. 1. This finding is in general accord with the results of Mitha et al. [12] and with the splash function of Anderson and Haff [1]. The correlation coefficient for the envelope is 0.71. The best fit line to the envelope has a gradient of 0.57, indicating that the ricochet speed is approximately 50—60% of the incident speed for a typical grain bed collision. This figure is in broad agreement with Anderson and Haff [1] who suggest the ricochet grain retained 40—50% of its impact speed. The data sample was too small to permit examinations through cross sections of the envelope.

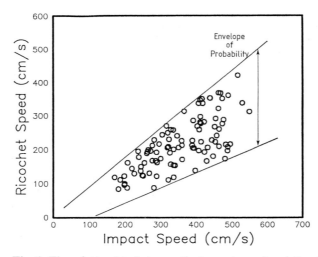

Fig. 1. The relationship between the impact speed and the ricochet speed for a medium grain impacting on a mixed bed (98 collisions represented)

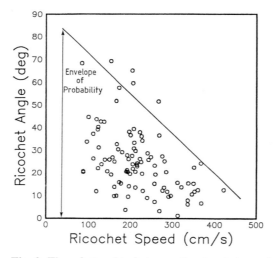

Fig. 2. The relationship between the ricochet speed and the ricochet angle (98 collisions represented)

---

[1] defined as the space representing the spread of data points

Mitha et al. [12] conducted collision experiments with steel ball bearings. They found that 6% of impacts did not rebound. Therefore the probability of a grain rebounding was taken to be 0.94 and independent of the impact speed. It is likely that this probability will be important in establishing the grain transport equilibrium. Unfortunately the high speed films [11] do not provide any confirmation of this probability.

The plot of ricochet angle against the ricochet velocity, shown in Fig. 2, displays a clear envelope showing a trend for the ricochet angle to decrease as the speed increases. The correlation coefficient is —0.46. The relationship is physically sensible as a sharply ascending grain might be expected to have had substantial contact with bed grains, losing a larger portion of its momentum. However a grain ricochetting at an oblique angle is more likely to have only glanced the surface, losing a smaller portion of its total momentum during the collision. Again the data sample was too sparse to permit an examination of cross sections of the envelope.

Therefore, knowing the impact speed of a grain the ricochet velocity may be found. Firstly, through recourse to the envelope shown in Fig. 1, to determine the ricochet speed and then, secondly, using this speed via Fig. 2, to fix the ricochet angle. In using both envelopes a random number, with uniform probability across the envelope, is used to determine the precise position in the envelope.

6 The ejected grains

The ejection of bed particles plays a crucial part in replacing saltation sequences which decay or terminate on collision. The momentum of the splashed grains must be transferred by the incident grain. Therefore a collision parameter of interest is the loss of momentum sustained by the rebounding grain [13].

Mitha et al. [12] noted that the number of ejections increased with the incident velocity. The data presented are in broad agreement with this observation (Fig. 3). However when the number of ejections per collision is plotted against the change in forward velocity (Fig. 4) then a more definite dependence is seen. This is physically sensible as the number of splashed grains might be expected to depend on the loss of momentum (princi-

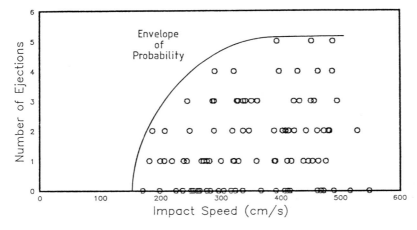

Fig. 3. The relationship between the impact speed and the number of grains ejected per collision (98 collisions represented)

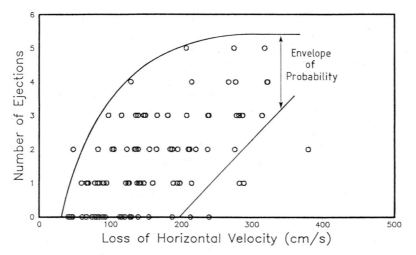

Fig. 4. The relationship between the loss of horizontal velocity at collision and the number of grains ejected per collision (98 collisions represented)

pally forward momentum). It should be noted that the exact size of an incident grain was not known in the experiment, hence it was not possible to calculate precisely the loss of forward momentum. Instead the loss of forward velocity was employed as an approximation. The momentum of a grain is dependent on the mass. The Balmedie sand used in the experiment is fairly compact in form so the mass of a grain is roughly cubically dependent on size. Therefore the largest grain in the medium fraction was almost 3 times heavier than the smallest, assuming uniform density. Nevertheless the approximation is the best available. It is expected that if the true values of forward momentum had been accessible then some of the scatter seen in Fig. 4 might be removed.

The envelope shown in Fig. 4 enables the loss of forward velocity of the rebounding grain to be partitioned between a number of ejections. A weak dependence was found between the ejection velocity of a grain and its angle of ascent Fig. 5. The dependence is similar to the relation between the ricochet speed and angle (Fig. 2). A qualitative explanation

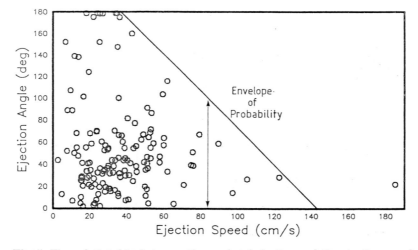

Fig. 5. The relationship between the angle of ejection and the ejection speed of a grain (147 grains represented)

of this is that the ejections with larger velocities will typically have been dislodged directly by the saltating grain (*primary dislodgement*) whereas those ejections with lesser velocities to have been dislodged by the jostling of neighbouring grains (*secondary dislodgement*). A primary dislodgement would characteristically receive a large forward impulse from the saltation causing a relatively large ejection velocity in a forward direction. A secondary dislodgement would receive a weaker impulse from neighbouring grains perhaps expelling the grain from the bed matrix in a more nearly vertical direction.

7 The air velocity modification

In 1941 Bagnold stated that "the sand profoundly alters the state of the wind". Yet, despite many studies, knowledge of the velocity profile within the saltation layer is restricted to time-averaged velocity measurements and rough calculations founded on these points. Frequently a "best fit" logarithmic profile has been forced to fit experimental measurements when the points themselves offer a variety of interpretations. Therefore there is a need for further study of the wind modification to which computer based models might contribute. Previous analytical models [3, 10] were limited in the range of trajectories they could calculate. In contrast to this the current generation of computer based models can calculate a spectrum of trajectories, ranging from hops of several centimeters to flights of over one meter. Their aggregated effect on the wind may be calculated and used to modify the velocity profile.

Owen [3] introduced the notion of grain borne shear stress. The grain borne shear stress at a height is defined as the force exerted by the grains on a column of air of unit base area extending from that height and upwards. Sørensen [4] coupled this notion to the assumption that the total shear stress was constant with height and equal to the sum of the air borne and grain borne shear stresses. This permits the calculation of the effective stress available to shear the air flow and thence the modified velocity profile may be calculated.

Ungar and Haff [10] and Werner [2] used a mixing length model to close the Navier Stokes equation in the forward direction of the wind.

$$F_x + \frac{d}{dy}\left(\varrho_a k^2 y^2 \left(\frac{du}{dy}\right)^2\right) = 0 \tag{3}$$

where F_x is the drag due to the grains

    ϱ_a is the density of air ($= 1.23\ \mathrm{kg/m^3}$)

    k is von Karman's constant ($= 0.4$)

    y is the height above the surface

    $\frac{du}{dy}$ is the velocity gradient

This equation represents a stationary wind, one that is neither accelerating nor decelerating, but it may be used iteratively to calculate the steady state wind. The drag force exerted by the grains is calculated from the trajectories. The wind is adjusted and a new group of trajectories is traced through the modified wind. The calculation is repeated until the effect of the trajectories and the wind are in equilibrium.

Alternatively, as in the present model, the left hand side of the equation may be used to calculate the decelerating force on the wind while the sand population moves towards equilibrium (a real time calculation) rather than be forced to close to zero. This calculation is achieved through the discretisation of the air flow with respect to height. Following the above equation each slice is assumed to have two forces acting on it: the drag of the sand grains (F_x) and the shearing force of the fluid $\left(d/dy(\varrho_a k^2 y^2 (du/dy)^2)\right)$.

$$\Delta u_a = \frac{1}{m_a} \int_{1}^{n} m_g \Delta u_g \, dn \qquad (4)$$

where Δu_a is the change in air velocity
       m_a is the mass of the air
       n is the number of grains
       m_g is the mass of a grain
       Δu_g is the change in velocity of a grain

The velocity gradient of the adjusted profile is calculated and from this the fluid shear stress is found as a function of height.

$$\tau_a = \varrho_a k^2 y^2 \left(\frac{du}{dy}\right)^2 \qquad (5)$$

where τ_a is the airborne shear stress
       ϱ_a is the density of air ($= 1.23 \text{ kg/m}^3$)
       k is von Karman's constant ($= 0.4$)
       y is the height above the surface
       $\dfrac{du}{dy}$ is the velocity gradient

This shear stress is assumed to act above and below a particular height discretisation and from this the shearing force acting over a finite time interval may be calculated and hence the change in velocity may be found.

The calculation demands one boundary condition. The particular one chosen is to set the shear stress well above the saltation layer to be equal to total shear stress at the bed. This is a steady state condition as the changes in total shear at the surface will take a finite time to influence the wind at a height well above the surface. Therefore the condition will be incorrect during the time the system is moving toward steady state but this discrepancy is thought to be small.

This approach is used to adjust the velocity profile in response to grain motion from the onset of saltation through to steady state and beyond. The adjustments occur at discrete time intervals, typically 0.01 s. A typical saltation trajectory has duration 0.1 s.

8 Results

A simulation was conducted for an initial shear velocity of 0.3 m/s with a uniform grain size of 300 µm. A steady state was attained after approximately 2.0 seconds. Figure 6 shows the wind profile at times $t = 0.0$ sec, $t = 1.0$ sec, $t = 2.0$ sec and $t = 10.0$ sec. Steady state saltation is demonstrated as the profiles show the wind changing little between $t = 0.0$ and $t = 1.0$ sec, decelerating rapidly between 1.0 and 2.0 sec and then remaining roughly constant from 2.0 sec to the end of the simulation at 10.0 sec. The plot of total mass flux against time (Fig. 11) shows a rapid increase in the transport rate between 1.0 and 2.0 sec until attainment of steady state.

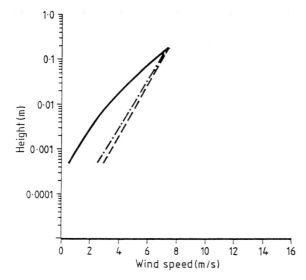

Fig. 6. Wind profiles at $t = 0.0$ sec (logarithmic, shown as dashed line), $t = 1.0$ sec (dot-dashed line), $t = 2.0$ sec (solid line) and $t = 10.0$ sec (solid line). The profiles at 2.0 sec and 10.0 sec are essentially similar indicating the attainment of steady state

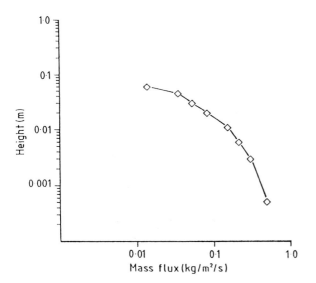

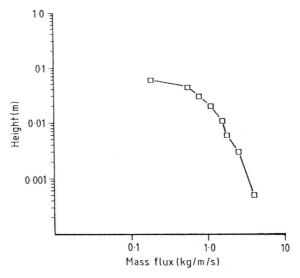

Fig. 7. The vertical mass flux profile shown on a logarithmic plane. The flux is calculated at 8 heights

Fig. 8. The horizontal mass flux profile shown on a logarithmic plane. The flux is calculated at the same 8 heights as Fig. 8

The horizontal and vertical mass flux profiles are shown plotted in a logarithmic plane in Figs. 7 and 8. These profiles conform to their expected shape with rapidly increasing flux approaching the bed and decaying to zero with increasing height from the bed. It is difficult to conclude that the vertical mass flux profile is linear in this plane although the horizontal mass flux may approximate to linearity.

The response of the calculated total mass flux to shear velocity is shown in Fig. 9. The model predicts that the value of total mass flux is proportional to the steady state shear velocity raised to the power of 2.97. This is in close agreement with Bagnold [1] who found this power to be 3.0. The magnitudes of the mass fluxes calculated are well within the range of those measured in experimental work [17].

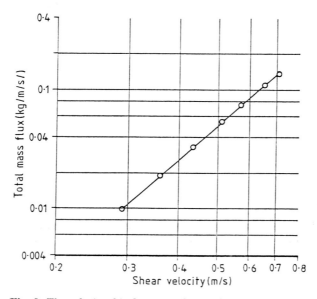

Fig. 9. The relationship between the total mass flux and the shear velocity calculated by the model. The model predicts the total mass flux to be proportional to the shear velocity raised to the power of 2.97

9 Discussion

Owen [3] suggested that, to the flow outside it, the saltation layer behaves as an increased aerodynamic roughness. This suggests that the shear stress well above the saltation layer will be changed by the presence of the saltating curtain. The effect of the roughness change will take some time to propagate upwards through the boundary layer so a final stationary value of shear stress will not be reached until some time after the attainment of the grain population steady state at roughly two seconds. This is illustrated in Fig. 10 which shows the value of shear velocity calculated from the wind profile at the top of the saltation layer against time and the value of the effective roughness against time. The shear velocity increases rapidly from its logarithmic value in the first two seconds and then attains a partial equilibrium from which it is gradually displaced by the response of the upper boundary layer to the change in effective roughness until the shear velocity tends to a final steady state. The slight decay in the shear velocity after the attainment of the primary steady state causes a corresponding, and more obvious decay in the total mass flux (Fig. 11) until the second and final equilibrium is reached. The decay in the mass flux is more noticeable because of its nonlinear dependence on the shear velocity. Therefore equilibrium sand movement is reached in two steps. Firstly, the sand transport rate must reach equilibrium with the wind in the saltation layer and secondly the boundary layer above the saltation layer must reach equilibrium with the roughness change due to saltation.

   The total shear stress in the saltation layer is not equal to the logarithmic shear stress before the onset of saltation. Rather, because of the increased roughness, shown in Fig. 10,

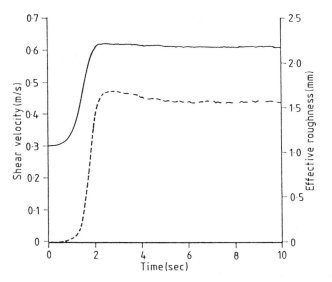

Fig. 10. The development of the shear velocity against time (solid line) and the change in effective roughness against time (dashed line) (both calculated 100 times/sec)

Fig. 11. The development of total mass flux against time. The curve attains a plateau indicating the arrival of steady state saltation. The total mass flux was calculated 100 times per second. The plot is unsmoothed

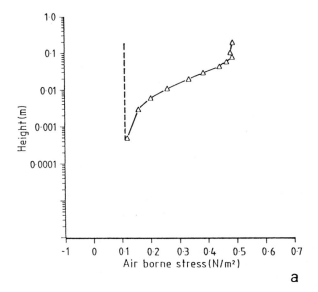

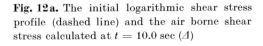

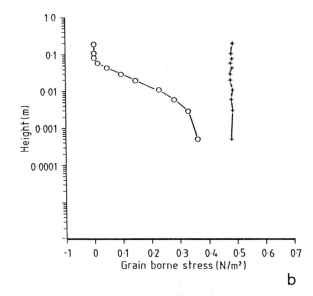

Fig. 12a. The initial logarithmic shear stress profile (dashed line) and the air borne shear stress calculated at $t = 10.0$ sec ($\triangle$)

Fig. 12b. The grain borne shear stress ($\circ$) calculated at $t = 10.0$ sec and the total shear stress ($+$) (the sum of the air and grain borne stresses)

it is several times larger. The grain borne shear stress is given by,

$$\tau_g(y) = \int_1^n m_g(u_2 - u_1) \, dn \tag{6}$$

where $\tau_g(y)$ is the grain borne stress at height, y

$\quad\quad m_g$ is the mass of a grain

$\quad\quad u_2$ is the horizontal grain velocity

$\quad\quad\quad$ during descent at height, y

$\quad\quad u_1$ is the horizontal grain velocity

$\quad\quad\quad$ during ascent at height, y

$\quad\quad n$ is the number of grains

Owen showed that the sum of the air and grain borne shear stresses should be constant and equal to the total shear stress. Figure 12a shows the initial logarithmic shear stress and the air borne shear stress [calculated by equ. (5) at $t = 10.0$ sec] and Fig. 12b shows the grain borne shear stress [calculated from equ. (6) at $t = 10.0$ sec] and the sum of the air borne and grain borne shear stresses. The sum is roughly constant as expected and the total stress in the saltation layer and above is in agreement with the steady state shear velocity shown in figure 10 ($\tau = \varrho U_*^2$).

This model predicts the airborne shear stress at the bed to be roughly equal to its initial logarithmic value suggesting that aerodynamic entrainment may be taking place at steady state. This is contrary to the conclusion of Anderson and Haff [1] who suggest that at steady state the shear stress at the bed has fallen below the threshold necessary to sustain aerodynamic entrainment. The principal difference in the two models appears to be that the Aberdeen model calculates the air borne and grain borne shear stresses explicitly whereas Anderson and Haff assume a total constant shear stress (equal to the initial logarithmic stress) and, through calculation of the grain borne shear stress, calculate the air borne shear stress implicitly by subtraction. This leads to an underestimate of the air borne shear stress at the surface. Willetts et al. [19] show that turbulent events are influential in entraining grains. It seems likely that these events will take place during saltation regardless of the bed shear stress. In fact the increase in shear velocity predicted by the model suggests the turbulent events may occur with greater intensity during saltation. However it may be seen from Fig. 6 that the model predicts the velocity near the surface to be close to zero. It is possible that this in itself may be sufficient to suppress aerodynamic entrainment as the pressure difference over the grains will be decreased. Therefore the model does not resolve the debate about the importance of aerodynamic entrainment at steady state, but rather points to the need for experimental work to support modelling investigations.

The above analysis contains an apparent contradiction. The Aberdeen model was developed on the assumption that impact entrainment was sufficient to establish steady state saltation and so no aerodynamic entrainment was incorporated in the model. The results appear to support this assumption yet the calculated wind profiles indicate that aerodynamic entrainment *may* take place at steady state. The saltating population may be considered as made up from two components,

$$N_a + N_i = N_t \tag{7}$$

where N_a is number of aerodynamically entrained grains

$\quad\quad N_i$ is the number of impact generated grains

$\quad\quad N_t$ is the total number of grains

If N_a is zero the saltating population N_t (set by the wind) will be entirely made up from impact entrained grains. In nature the relative magnitudes of N_a and N_i will depend on factors such as wind characteristics, grain size and shape, [20] whereas, with regard to numerical models the relative magnitudes of N_a and N_i at steady state will depend strongly on the rules used to simulate aerodynamic entrainment. Preliminary experiments suggest these rules will not be simple [19] and that it will be a difficult task to establish the role of aerodynamic entrainment at steady state. Therefore the overall results from the model remain valid but suggest that immediate attention should be given to incorporation of a rule to model aerodynamic entrainment accurately.

10 Further work

Preliminary numerical experiments have been conducted using the saltation model to study ripple development. The model can calculate the impact rate and the exchange of momentum with the bed which are critical to ripple development. Willetts and Rice [18] extended their experimental study of collision to include inclined beds. These results may be incorporated into a modified splash function dependent upon the bed slope at impact. Inclusion of this splash function in the model enable the calculation to take place over an undulating surface if the effect of the surface perturbations on the wind is neglected. The saltation cloud would be periodically recalculated in response to a changing bed until the entire system attains equilibrium.

Fletcher [19] noted that the saltation of large grains was responsible for the erosion of fine dust in wind tunnel experiments. Using this observation, the model may be used to estimate dust emissions due to saltation when coupled with a model of the dust entrainment process from the surface.

11 Conclusions

The experimental data used to develop the splash function shows some clear relationships between collision parameters. The ricochet speed is shown to be strongly dependent on the impact speed and the ricochet angle to be weakly dependent on the ricochet speed. The loss of forward velocity at collision is influential in determining the number of ejected grains. These grains are recognised as important in replacing saltation sequences which terminate or decay.

The simulation results are in close agreement with reported experimental work. The wind profile is calculated from consideration of the air/grain momentum exchange and the action of the air borne shear stress at different heights. The calculated wind profile is similar in shape to experimentally measured profiles and resembles the calculated profile of Anderson and Haff [1] although the calculation procedure is different. The relationship between total mass flux and shear velocity is approximately cubic. Moreover the flux profiles are realistic. Thus preliminary results encourage confidence in basic features of the model. However initial results indicate the shear stress at the bed may be high enough to sustain some aerodynamic entrainment at steady state. An appeal is made for more experimental work to resolve this debate.

Acknowledgements

NERC have provided the financial support for this study. The authors are especially grateful to M. A. Rice for her careful work in analysing the collision films. We also thank R. S. Anderson and N. O. Jensen for helpful advice.

References

[1] Anderson, R. S., Haff, P. K.: Simulation of eolian saltation. Science **241**, 820—823 (1988).

[2] Werner, B. T.: A steady-state model of wind blown sand transport. J. Geol. **98**, 1—17 (1990).

[3] Owen, P. R.: Saltation of uniform sand grains in air. J. Fluid. Mech. **20**, 225—242 (1964).

[4] Sørensen, M.: Estimation of some eolian saltation transport parameters from transport rate profiles. Proc. Int. Wkshp. Physics of Blown Sand **1**, 141—190 (1985).

[5] White, B. R., Schulz, J. C.: Magnus effect in saltation. J. Fluid Mech. **81**, 497—512 (1977).

[6] Hunt, J. C. R., Nalpanis, P.: Saltating and suspended particles over flat and sloping surfaces I. Modelling concepts. Proc. Int. Wkshp. Physics of Blown Sand **1**, 9—36 (1985).

[7] Morsi, S. A., Alexander, A. J.: An investigation of particle trajectories in two-phase flow systems. J. Fluid Mech. **55**, 193—208 (1972).

[8] Schiller, L., Nauman, A.: Z. Ver. Dent. Ing. **77**, 318 (1933).

[9] White, F.: Viscous flow. New York: McGraw-Hill 1974.

[10] Ungar, J. E., Haff, P. K.: Steady-state saltation in air. Sedimentology, **34**, 289—299 (1987).

[11] Willetts, B. B., Rice, M. A.: Inter-saltation collisions. Proc. Int. Wkshp. Physics of Blown Sand **1**, 83—100 (1985).

[12] Mitha, S., Tran, M. Q., Werner, B. T., Haff, P. K.: The grain bed impact process in aeolian saltation. Acta Mech. **63**, 267—278 (1986).

[13] Willetts, B. B., Rice, M. A.: Collision in aeolian transport: the saltation/creep link. In: Nickling, W. G. (ed.) Aeolian geomorphology. Allen & Unwin, pp. 1—17.

[14] Owen, P. R.: The physics of sand movement. Lecture Notes, Wkshp on physics of desertification, Trieste, (1980).

[15] Gerety, K. M.: Problems with determination of U_* from wind velocity profiles in experiments with saltation. Proc. Int. Wkshp. Physics of Blown Sand **2**, 271—300 (1985).

[16] Bagnold, R. A.: The physics of wind blown sand and desert dunes. London: Chapman and Hall 1973.

[17] Rasmussen, K. R., Mikkelsen, H. E.: Aeolian transport in a boundary layer wind tunnel. Geoskrifter Nr. 29, Geological Institute, Aarhus University (1988).

[18] Willetts, B. B., Rice, M. A.: Collision of quartz grains with a sand bed: the influence of incident angle. Earth Surface Processes and Landforms **14**, 719—730 (1989).

[19] Willetts, B. B., McEwan, I. K., Rice, M. A.: Initiation of motion of quartz sand grains (this volume).

[20] Rice, M. A.: Grain shape effects on aeolian sediment transport (this volume).

[21] Fletcher. B.: The erosion of dust by an airflow. J. Phys. D: Appl. Phys. **9**, 913—924 (1976).

Authors' address: I. K. McEwan and B. B. Willetts, Department of Engineering, University of Aberdeen, Kings College, Aberdeen AB92UE, United Kingdom.

Acta Mechanica (1991) [Suppl] 1: 67—81

An analytic model of wind-blown sand transport

M. Sørensen, Aarhus, Denmark

Summary. An analytic steady-state model of wind-blown sand transport is developed. The basic physical ideas are as in the current numerical models. In order to obtain explicit analytic expressions, approximations are made in all of the three subprocesses constituting the model: the trajectory model, the wind modification model, and the grain-bed collision model. Explicit formulae are derived for the transport rate and for the flux of grains from the bed into the air. The transport rate is found to be an approximately cubic function of the friction velocity. A number of equilibrium equations are important in the theory developed in the paper. These equations are derived and thoroughly discussed in the setting of a general model of aeolian sand transport and are generally applicable. Possible refinements of the theory are briefly discussed.

1 Introduction

In recent years there has been very considerable progress in the development of mathematical models for aeolian saltation, [3], [4], [8], [17]. This development is mainly due to important progress in our conceptual understanding of the process of aeolian sand transport; for a review see [5]. Because of the coupling of the several mechanisms active in determining the equilibrium state of the transport process, the models mentioned above do not give explicit analytic expressions for the transport rate and other quantities of interest. Such quantities must be determined by computer simulations. These models reproduce several empirically well-established features of aeolian saltation.

There is, however, a need to supplement the computer simulations by models based on the current understanding of the process of aeolian sand transport, but simple enough to allow explicit analytic calculation. Here a first attempt in this direction is presented. In particular, we seek formulae for the transport rate and for the flux of sand from the bed into the air as functions of the friction speed. Much of the basic physics is the same as in the above-mentioned models, but by introducing approximations, some of the mechanisms of aeolian saltation are decoupled. The trajectory models used are based on approximations introduced by P. R. Owen [11], [12]. The same grain-wind interaction model is used as in [15]. It is argued that this is an approximate version of R. S. Anderson's model [1]. The bed-grain interaction is based on the splash laws used by Anderson and Haff [3], [4] and Werner [17]. Only sand with a single grain size is considered.

In Section 2 two analytic trajectory models are discussed. In the first model the drag on a grain is approximated by a linear function of the relative velocity between air and grain. We shall mainly use this trajectory model, but also a second model is needed, which is obtained by imposing the further simplification that the wind speed is independent of height. In Section 3 conditions for stability of the sand transport are discussed in detail. The equilibrium state of the transport process is found by equating the mean number of grains ejected by an impinging grain to an experimental value. We assume that the probability distribution of the launch velocity vector is independent of the wind speed and find the mass flux from the bed into the air needed to retard the wind enough that the equilibrium equation is satisfied. From this flux follows an explicit expression for the transport rate. For a full specification of the model, we still need the probability distribution of the launch velocity of the grains. In order to find the equilibrium distribution of the launch velocity, some probabilistic aspects of the splash law are considered in Section 4. In particular, expressions are derived for the fraction of grains in transit that are saltating and the fraction of reptating grains. The saltating grains are the high-energy grains jumping along, while the reptating grains are the low-energy ejecta. An equilibrium gamma distribution is found using a splash law based on [3], [17] and [19].

The self-limiting mechanism is rather crudely modelled in the present paper, mainly because of incomplete knowledge about the bed-grain interaction. It is, however, possible to incorporate more intricate self-limiting mechanisms in the model as more information about the bed-grain interaction becomes available. This aspect is discussed in Section 5.

2 Analytic trajectory models

Crucial parts of the analytic theory of saltation presented in this paper are two approximative models of the trajectories of saltating and reptating grains. These trajectory models, which were proposed by Owen [11], [12], capture much of the essence of aeolian grain motion.

The equations of motion used in most trajectory models are

$$\ddot{x} = H(v)\left(U(y) - \dot{x}\right)$$
$$\ddot{y} + g + H(v)\,\dot{y} = 0,$$
(2.1)

where x is the horizontal distance from the starting point of the trajectory to the present position of the grain, and y is the altitude of the grain. Further, g is the gravitational acceleration, $U(y)$ is the time-averaged wind speed at height y and v is the relative velocity between the grain and the air. The function $H(v)$ is given by

$$H(v) = D(v)/mv,$$
(2.2)

where $D(v)$ is the drag exerted on the grain at the relative velocity v, and m denotes the mass of the grain. The drag is usually calculated as the drag on an aerodynamically equivalent sphere in the sense of Bagnold [6]. The effects of fluid lift and of the turbulent eddies are not taken into account.

There are two couplings between the equations (2.1), through $H(v)$ and through $U(y)$, which render an analytic solution impossible. Owen [11] proposed to get around one of these problems by using a linear approximation to the drag:

$$D(v) = \delta v$$
(2.3)

where δ should be chosen such that (2.3) gives the correct drag on an aerodynamically equivalent sphere at a typical relative velocity v_*. The quality of this approximation was discussed in [12]. Calculations using the trajectory model in [7] with initial values from [19] indicates that for friction velocities in the range 45—60 cm/s the value $v_* = 235$ cm/s is reasonable for saltating grains of size 300 μm. Under the same conditions $v_* = 180$ cm/s can be used for reptating grains.

Using (2.3), we find that $H(v) = \delta/m$ so that the y-equation is decoupled from the x-equation:

$$t_* \ddot{y} + v_f + \dot{y} = 0. \tag{2.4}$$

The quantity $t_* = m/\delta$ has the dimension of time. As discussed in [2] and [12], t_* can be interpreted as a reponse time of the grain to changes in the wind speed. To facilitate the notation we introduce the quantity $v_f = g t_*$ with the dimension of a velocity. The velocity v_f is a decreasing function of v_*. For grains of size 300 μm we find $v_f = 232$ cm/s for $v_* = 235$ cm/s and $v_f = 262$ cm/s for $v_* = 180$ cm/s.

The solution to (2.4) is

$$y(t) = t_*(v_f + v_2{}^0)\,(1 - e^{-t/t_*}) - v_f t \tag{2.5}$$

$$\dot{y}(t) = v_2{}^0 e^{-t/t_*} - v_f(1 - e^{-t/t_*}), \tag{2.6}$$

where $v_2{}^0$ is the vertical component of the launch velocity vector. The duration of the jump t_i is the solution of $y(t_i) = 0$. It follows that

$$v_2{}^0 = v_f \left\{ \frac{t_i/t_*}{1 - \exp(-t_i/t_*)} - 1 \right\} \tag{2.7}$$

and

$$\dot{y}(t_i) = v_2{}^0 - g t_i, \tag{2.8}$$

whereas an explicit expression for t_i as a function of $v_2{}^0$ can not be given. Occasionally, we shall need an explicit expression. In such cases we use the linear approximation

$$t_i \doteq 1.75 v_2{}^0/g, \tag{2.9}$$

which can be compared to the zero-drag result, $t_i = 2 v_2{}^0/g$. The approximation (2.9) deviates by at most 8% from the jump times given by (2.7) provided $v_2{}^0/v_f$ is in the range 0.2—0.9, which happens with a high probability. In the range 0.1—1 the accuracy is better than 10%. Based on the observed lift-off velocities reported in [19], we expect $v_2{}^0/v_f$ to be in the range 0.2—1.4 for saltating grains and in the range 0—0.3 for reptating grains.

Formula (2.9) implies that the vertical component of the impact velocity is approximated by

$$\dot{y}(t_i) \doteq -0.75 v_2{}^0. \tag{2.10}$$

Because of the deceleration due to the drag in the vertical direction, the grain does not return to the bed with the same vertical velocity component as it started with.

The time $\bar{t}$ at which the grain is at the top of its trajectory is the solution of $\dot{y}(t) = 0$, i.e. by (2.6), $\bar{t} = t_* \ln(1 + v_2{}^0/v_f)$. Thus the height of the trajectory is

$$y(\bar{t}) = t_* \{ v_2{}^0 - v_f \ln(1 + v_2{}^0/v_f) \}. \tag{2.11}$$

Note that by a second order Taylor expansion of the logarithm, the zero-drag result $y(\bar{t}) = \frac{1}{2}(v_2{}^0)^2/g$ is obtained from (2.11). Hence the difference between (2.11) and the zero-drag formula is of the order $0\big((v_2{}^0/v_f)^3\big)$. The range of likely values of $v_2{}^0/v_f$ was discussed above.

The x-equation has not by the linear drag approximation been decoupled from the y-equation. The x-equation

$$t_* \ddot{x} = U(y) - \dot{x}$$

can none the less be solved:

$$x(t) = \int_0^t [1 - e^{-(t-s)/t_*}]\, U\big(y(s)\big)\, ds + t_* v_1^0 (1 - e^{-t/t_*}) \tag{2.12}$$

$$\dot{x}(t) = t_*^{-1} \int_0^t e^{-(t-s)/t_*}\, U\big(y(s)\big)\, ds + v_1^0 e^{-t/t_*}, \tag{2.13}$$

where v_1^0 is the horizontal component of the launch velocity vector. At some points we shall need a simpler model for $x(t)$. This can be obtained by imposing the further simplification that $U(y)$ is constant. This approximation, which was proposed by Owen [12], more seriously damages the realism of the trajectory calculations than did the linear drag approximation (2.3). Fortunately, it affects only the horizontal motion of the grain. The value of U should be chosen as the wind speed at the height at which most of the grain-wind interaction takes place.

The specification $U(y) = \overline{U}$ decouples the equations (2.1) completely and we find

$$x(t) = t_*(\overline{U} - v_1^0)\, (e^{-t/t_*} - 1) + \overline{U}t \tag{2.14}$$

$$\dot{x}(t) = \overline{U}(1 - e^{-t/t_*}) + v_1^0 e^{-t/t_*}. \tag{2.15}$$

Note that

$$\frac{\dot{x}(t_i) - v_1^0}{x(t_i)} = \frac{g(\overline{U} - v_1^0)}{\overline{U}v_2^0 + v_1^0 v_f},$$

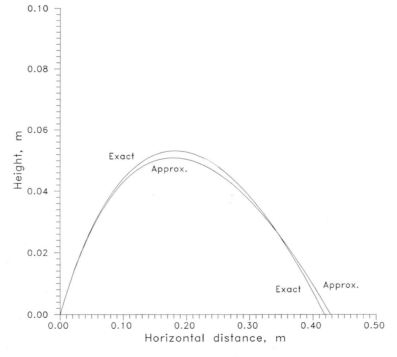

Fig. 1. The effect of the drag approximation (2.3) on a typical saltation trajectory. The launch angle is $35°$ and the launch speed is 200 cm/s. The wind speed is 350 cm/s at all heights. One trajectory (exact) was calculated numerically, the other (approx.) is given by (2.5) and (2.14)

which for $v_1{}^0 = 0$ equals $g/v_2{}^0$. Bagnold [6, p. 65] found the result for $v_1{}^0 = 0$ to hold very closely over a wide range of $v_2{}^0$ values.

The effect of the drag approximation (2.3) on a typical trajectory is shown in Fig. 1. Here two trajectories are compared for a grain launched at an angle of $35°$ and with a speed of 200 cm/s. The wind speed is assumed the same at all heights. One trajectory (exact) was calculated numerically using the trajectory model in [7] while the other (approx.) is given by (2.5) and (2.14).

3 Equilibrium saltation

In this section we derive explicit formulae for the transport rate and for the flux of grains from the bed into the air. We begin by deriving and discussing in some detail two basic conditions for equilibrium. One of these conditions states that when a saltating grain hits the bed, the average number of saltating grains that result from the collision must equal one. Our next step will be a model for how the grains in transport modify the wind profile. This model uses an eddy viscosity based on the concept of grain borne shear stress. We find the modified wind profile for which an equilibrium equation is satisfied. This gives us an expression for the equilibrium flux of grains from the bed into the air, namely the flux needed to obtain the wind profile modification. Based on the theory developed in the paper a formula for the transport rate can now be derived.

We have to be precise in what we mean by a saltating grain. We call a grain that has rebounded at least once a *saltating* grain, while the motion of a grain that has been ejected by an impinging grain will be called *reptation*. For a more thorough discussion of these transport modes see [5].

Let $\bar{n}$ denote the average number of grains ejected by an impinging saltating grain. The average is over all possible impact velocities as well as over all possible outcomes of the collision. By N and αN we denote the average number of saltators and reptators that impact on unit area of the bed per second. Thus α is the ratio of the number of reptators to the number of saltators in the saltation cloud. The number of reptators (ejecta) that leaves unit area per second is $N\bar{n} + \alpha N\delta$, where δ is the average number of grains ejected when a reptator hits the bed. At equilibrium the number of reptators in the saltation cloud is constant, so it must hold that

$$\alpha N = N\bar{n} + \alpha N\delta,$$

which is equivalent to

$$\alpha = \bar{n}/(1 - \delta). \tag{3.1}$$

If we denote the fraction of saltators in the saltation cloud by π_s, we find that

$$\pi_s = \frac{1 - \delta}{1 + \bar{n} - \delta}. \tag{3.2}$$

The next equilibrium equation is expressed in terms of $\bar{n}$, p_s and p_r. where p_s and p_r denote the probability that a saltating grain and a reptating grain, respectively, rebounds when hitting the bed. The number of grains that rebound off unit area of the bed per second is $Np_s + \alpha Np_r$. Therefore, to ensure a constant fraction of saltators among the grains in motion, it must hold that

$$N = Np_s + \alpha Np_r.$$

By inserting (3.1) we find the equilibrium condition

$$p_s + \bar{n}p_r/(1 - \delta) = 1. \tag{3.3}$$

If instead we insert the identity $\alpha = (1 - \pi_s)/\pi_s$, we obtain

$$\pi_s = p_r/(1 - p_s + p_r).$$

Note that this relation does not involve δ. The number δ is very small, and if we set $\delta = 0$, we obtain the equilibrium equations

$$\pi_s = \frac{1}{1 + \bar{n}} \tag{3.4}$$

and

$$p_s + \bar{n}p_r = 1. \tag{3.5}$$

We shall in this section mainly focus on (3.5). Obviously, the three quantities in (3.5) depend on characteristics of the sand such as size-distribution, grain shape and elasticity. Presumably they also depend on the wind speed. To discuss this question more thoroughly, let $p(R|\boldsymbol{v}^i)$ denote the probability of rebound for an impinging grain with impact velocity vector $\boldsymbol{v}^i$. Let similarly $n(\boldsymbol{v}^i)$ be the average number of grains ejected given the impact velocity. Then

$$p_s = \int p\big(R|I(\boldsymbol{v}^0)\big) f_s(\boldsymbol{v}^0)\, d\boldsymbol{v}^0, \tag{3.6}$$

$$p_r = \int p\big(R|I(\boldsymbol{v}^0)\big) f_r(\boldsymbol{v}^0)\, d\boldsymbol{v}^0 \tag{3.7}$$

and

$$\bar{n} = \int n\big(I(\boldsymbol{v}^0)\big) f_s(\boldsymbol{v}^0)\, d\boldsymbol{v}^0, \tag{3.8}$$

where f_s and f_r are the probability densities of the launch velocity for saltating and reptating grains respectively, and where I is the function relating the impact velocity after a jump to the velocity $\boldsymbol{v}^0$ with which the grain was launched. The function I is determined by the trajectory model. The discussion here is phrased to cover any deterministic trajectory model. We see from (3.6)—(3.8) that $\bar{n}$, p_s and p_r can depend on the wind speed through I, f_s and f_r. Werner [17] found that the probability distribution of the launch velocity of an ejected grain can be assumed independent of the velocity of the impinging grain that dislodges it. From this it follows that f_r does not depend on the wind speed. In [7] it was found that f_s does not depend much on the wind speed except at low wind speeds. In the next section conditions will be given under which f_s is independent of the wind speed. The main source of dependence of p_s and p_r on the wind speed is the function $\boldsymbol{v}^0 \to p\big(R|I(\boldsymbol{v}^0)\big)$, where I is specified by a trajectory model. We see that a crucial quantity for determining the equilibrium state of aeolian saltation is the conditional probability $p(R|\boldsymbol{v}^i)$. Unfortunately, this probability has so far received very little attention in the saltation literature.

The average number $n(\boldsymbol{v}^i)$ of grains ejected by a grain hitting the sand bed with velocity $\boldsymbol{v}^i$ has attracted much more interest. Anderson and Haff [3], [4] found that $n(\boldsymbol{v}^i)$ is proportional to $|\boldsymbol{v}^i|$. Here we assume that at equilibrium all saltating grains hit the bed at an angle of $12°$. Under this assumption the expression in [3] becomes

$$n(\boldsymbol{v}^i) = 0.0061 v_1{}^i. \tag{3.9}$$

Here and elsewhere we use the *cgs* system of units.

If we use (2.6) and (2.15) to define the function I, and work under the assumption that f_s and f_r are independent of the wind speed, then (3.5) gives an equation for $\bar{U}$, the typical wind speed used in (2.15). The value of $\bar{U}$ which solves (3.5) is the same at all friction velocities. We interpret $\bar{U}$ as the wind speed at a typical height $\bar{y}$, and set $\bar{y}$ equal to 0.85 times the mean jump height of the grains in motion, i.e. $\bar{y} = 2.2$ cm (see Section 4). Because $\bar{U}$ does not depend on the friction speed, it follows from (3.6)—(3.8) that p_r, p_s and $\bar{n}$ have the same value at all friction speeds. Willetts and Rice [19] found the value $\bar{n} = 1.75$ for quartz grains of typical size 300 μm.

We will now find the value of $\bar{U}$ corresponding to $\bar{n} = 1.75$. To do this, we take the mean value of (3.9) and find that $1.75 = 0.0061 E_s(v_1{}^i)$. Next we insert into this equation the expression for $E_s(v_1{}^i)$ obtained by taking the mean value of the formula derived from (2.15) for $v_1{}^i = \dot{x}(t_i)$. In this way we obtain

$$\bar{U} = [2.87 - E_s(v_1{}^0 e^{-t_i/t_*})]/[1 - E_s(e^{-t_i/t_*})]. \tag{3.10}$$

Here E_s denotes expectation with respect to the launch velocity distribution of a rebounding grain. Using the equilibrium launch velocity distribution derived in the next section, we find that $\bar{U} = 333$ cm/s. This value of the wind speed at height 2.2 cm is not much above what has been found in field experiments [14], but somewhat lower than what has been measured in wind tunnels [11], [13]. The equilibrium value $\bar{U}$ of the wind speed at $\bar{y}$ can not be attained if the friction velocity is below

$$U_*{}^c = k\bar{U}/\ln(\bar{y}/y_0), \tag{3.11}$$

where $k = 0.4$ is von Karman's constant and y_0 is the height at which the wind speed vanishes. Our interpretation of this is that for $U_* < U_*{}^c$ the wind is too feeble to maintain the sand transport. Thus the critical friction velocity $U_*{}^c$ is closely related to the *impact threshold*. We set $y_0 = 0.01$ cm corresponding to small ripples. With $\bar{y} = 2.2$ cm as discussed above, we find $U_*{}^c = 25$ cm/s.

The use of $\bar{n} = 1.75$ is not quite satisfactory since the sand transport in Willetts and Rice's experiment was not in equilibrium with the wind. Moreover, the sand used in the experiment had a rather broad spectrum of sizes. It is, however, the best we can do with the information available at present. A more refined theory with $\bar{U}$ depending on the friction speed can be obtained by using (2.13) for $v_1{}^i$ instead of (2.15). This, however, requires information about $p(R|\boldsymbol{v}^i)$. We shall discuss this refinement in Section 5.

We now seek, for each value of the friction speed U_*, the mass flux ϕ of sand from the bed into the air needed to fix the wind speed at height $\bar{y}$ at the value $\bar{U}$ given by (3.10). The grain borne shear stress at height y is

$$T(y) = \phi E\big(\Delta\dot{x}(y)\big), \tag{3.12}$$

see [15], where $\Delta\dot{x}(y)$ denotes the change in the horizontal speed while the grain is above the level y, and where E denotes expectation with respect to the probability distribution of the launch velocity, irrespective of whether the grain is ejected or rebounding. Consider a trajectory with maximum height larger than y, and let $t_1(y)$ and $t_2(y)$ denote the first and the second time, respectively, the grain is at height y. Then by (2.15)

$$E\big(\Delta\dot{x}(y)\big) = E[(\bar{U} - v_1{}^0)\,(e^{-t_1(y)/t_*} - e^{-t_2(y)/t_*});\, v_2{}^0 > v_2(y)]. \tag{3.13}$$

The expectation is calculated for values of $v_2{}^0$ only that are larger than $v_2(y)$, the smallest vertical launch speed for which the level y is attained. The value of $v_2(y)$ is found from (2.11).

The wind profile $U(y)$ in the saltation layer can be calculated by specifying an eddy viscosity as discussed by several authors [11], [15], [1], [16], [17]. Anderson [1] used the eddy viscosity

$$\nu(y) = ky(U_*{}^2 - T(y)/\varrho)^{1/2}, \tag{3.14}$$

where k is von Karman's constant, ϱ is the density of the air and U_* is the friction speed in the grain-free logarithmic layer above the saltation layer, see also [4]. Note that (3.14) differs from the eddy viscosity in a grain-free logarithmic layer in that U_* has been replaced by a corrected friction speed taking into account the reduction of the air borne shear stress. By (3.14)

$$\frac{dU}{dy} = \frac{U_*}{ky}\,[1 - T(y)/(\varrho U_*{}^2)]^{1/2}. \tag{3.15}$$

If we, somewhat crudely, approximate the function $\sqrt{1-x}$ by $1-x$, we obtain the equation

$$\frac{dU}{dy} = \frac{U_*}{ky} - \frac{T(y)}{ky\varrho U_*}. \tag{3.16}$$

This is done to obtain an equation that is linear in ϕ, which will allow us to obtain an explicit expression for ϕ as a function of U_*. Equation (3.16) was derived in [15] using the eddy viscosity $\nu(y) = kyU_*$. The reason for choosing the approximation $1-x$ rather than a Taylor expansion is that it is important to ensure the right magnitude for all values of x between zero and one. From (3.16) it follows that

$$U(y) = (U_*/k) \ln (y/y_0) - \frac{\phi}{k\varrho U_*} \int_{y_0}^{y} E\big(\Delta\dot{x}(z)\big)\, z^{-1}\, dz, \tag{3.17}$$

where y_0 is the height at which the wind speed vanishes. As in e.g. [17] we use a Nikuradse-type condition and set y_0 equal to 0.01 cm corresponding to small ripples.

Now we can find ϕ from (3.17) by equating $U(\bar{y})$ to $\bar{U}$:

$$\phi = \frac{\varrho U_*[U_* \ln (\bar{y}/y_0) - k\bar{U}]}{\displaystyle\int_{y_0}^{\bar{y}} E\big(\Delta\dot{x}(z)\big)\, z^{-1}\, dz}. \tag{3.18}$$

Note that ϕ is proportional to $\varrho U_*(U_* - U_*{}^c)$ if we use (3.11) for $U_*{}^c$. With the equilibrium launch distribution derived in Section 4 and with y_0 and $\bar{y}$ as above we find that

$$\phi = 0.022\varrho U_*(U_* - U_*{}^c). \tag{3.19}$$

This formula yields values of ϕ larger than what was estimated from observed transport rate profiles in [7] and [15]. This is typical of the present generation of self-limiting saltation models. At $U_* = 50$ cm/s a ϕ value of 0.3 g cm^{-2} min^{-1} was estimated in [7] and [15]. This value found some support in [20] where the value 0.1 g cm^{-2} min^{-1} was derived from a very different type of observations. At the same wind speed formula (3.19) gives the ϕ value 2.0 g cm^{-2} min^{-1}, while Anderson and Haff's [4] model gives 1.3 g cm^{-2} min^{-1}. McEwan and Willetts [8] found the value 8.8 g cm^{-2} min^{-1} at $U_* = 40$ cm/s. In the papers cited here the typical grain size was 300 μm, except in [4] where it was 250 μm. A possible explanation of the discrepancy between theory and observations is that in the experiments on which [7] and [15] are based a considerable fraction of the reptating population was not observed. Another possible explanation is that other self-limiting mechanisms than the wind speed modification are active in nature. We shall briefly return to this point in the last section.

Using (3.12) and (3.19) we find that the ratio between grain borne shear stress at the bed and total shear stress is $1.25(1 - U_*^c/U_*)$. Note that for $U_* > 5U_*^c$ unrealistic ratios larger than one are obtained. Thus for U_* larger than 125 cm/s the model does not work. This is not a problem from a practical point of view. When U_* is near U_*^c, the air borne shear stress at the bed is close to that given by U_*^c, as hypothesised by Owen [11]. As U_* increases, the air borne shear stress at the bed declines. This accords with findings in [17].

By inserting (3.19) in (3.17) we obtain

$$U(y) = 2.5U_* \left\{ \ln(y/y_0) - 0.022(1 - U_*^c/U_*) \int_{y_0}^{y} E(\Delta\dot{x}(z)) z^{-1} dz \right\}. \qquad (3.20)$$

In the grain-free layer above the saltation layer (3.20) becomes logarithmic. The apparent zero-point calculated from this logarithmic profile increases with U_* as expected from empirial findings, [11]. The exact functional dependence of the apparent zero-point does not accord with Owen's empirical formula, [11]. Note, that (3.20) has a fix-point at $y = \bar{y}$. Note also that when y_0 is chosen in a reasonable range, the exact value of y_0 has little influence on the wind speeds calculated by (3.20). In fact, the wind speed in the saltation layer given by (3.20) vanishes well above the height y_0.

When the probability distribution of the launch velocity for saltating and reptating sand has been specified we have a complete model of aeolian sand transport. This problem is considered in the next section. With a complete model we can calculate quantities of practical interest such as the transport rate which equals the product of the flux ϕ and the mean hop length $E(x(t_i))$, see e.g. [15]. Using (2.12) we find that

$$\text{transport rate} = 2.5U_* \phi E \left\{ \int_0^{t_i} [1 - e^{-(t_i-t)/t_*}] \ln(y(t)/y_0) \, dt \right\}$$

$$- \frac{2.5\phi^2}{\varrho U_*} E \left\{ \int_0^{t_i} [1 - e^{-(t_i-t)/t_*}] \int_{y_0}^{y(t)} E(\Delta\dot{x}(z)) z^{-1} dz \, dt \right\} + \phi t_* E\{v_z^0(1 - e^{-t_i/t_*})\}, \qquad (3.21)$$

with ϕ given by (3.18). Since ϕ goes approximately as U_*^2, the transport rate is thus found to increase approximately as the friction velocity cubed. With the launch velocity distribution found in the next section, (3.21) becomes

$$\text{transport rate} = 0.0014\varrho U_*(U_* - U_*^c)(U_* + 7.6U_*^c + 205). \qquad (3.22)$$

The ratio $R = \text{transport rate}/U_*^3$ expresses the departure from exact proportionality to U_*^3. With the transport rate given by (3.22), R increases with U_* up to 53 cm/s. Above this U_* value R decreases towards a constant value for increasing U_*. A similar behaviour of R has been found experimentally, see [13]. Quantitatively, however, formula (3.22) yields transport rates that are about 3 times those predicted by Bagnold's formula [6] and about 3.5 times the experimental transport rates in [13]. Also McEwan and Willetts' model [8] gives transport rates that are significantly larger than what has been found empirically, whereas the models of Werner [17] and Anderson and Haff [4] give transport rates of a magnitude in reasonable accordance with experimental values.

4 The splash law and the launch velocity distribution

The splash law is the probability law that, given the velocity of an impinging grain, specifies the probability with which the grain rebounds, the distribution of its take-off velocity, the distribution of the number of grains ejected and the distribution of their launch velocities. The splash law has so far only been partly studied and is still a subject of considerable research effort.

Our aim in this section is to determine equilibrium probability density functions of the launch velocity vector for saltating and reptating grains. As in the preceeding section we denote these density functions by $f_s(\boldsymbol{v})$ and $f_r(\boldsymbol{v})$. First we derive general equilibrium equations. Next we obtain a more manageable equation by using a splash law of a suitable form. Finally, we consider the case where f_s is a gamma-density.

To get started, we need to determine the fraction π_s of saltating grains among the grains in motion. To do this let, as in Section 3, $p(R|\boldsymbol{v}^i)$ denote the probability of rebound for a grain that hits the bed with velocity $\boldsymbol{v}^i$, and let $p(n|\boldsymbol{v}^i)$ be the probability that n grains are ejected given $\boldsymbol{v}^i$.

The probability $q_0(\boldsymbol{v}^i)$ that no grains leave the surface after an impact event with impact velocity $\boldsymbol{v}^i$ is given by $q_0(\boldsymbol{v}^i) = p(0|\boldsymbol{v}^i)\big(1 - p(R|\boldsymbol{v}^i)\big)$ provided the events of rebound and ejection are independent. Under the same assumption, the fraction $p_s(\boldsymbol{v}^i)$ of saltating grains resulting from an impact with velocity $\boldsymbol{v}^i$, given that at least one grain leaves the bed, is given by

$$q_+(\boldsymbol{v}^i)\, p_s(\boldsymbol{v}^i) = \sum_{n=0}^{\infty} (n+1)^{-1}\, p(n|\boldsymbol{v}^i)\, p(R|\boldsymbol{v}^i) = p(R|\boldsymbol{v}^i)\, E\big((1+N)^{-1}|\boldsymbol{v}^i\big). \tag{4.1}$$

Here $q_+(\boldsymbol{v}^i) = 1 - q_0(\boldsymbol{v}^i)$ is the probability that at least one grain leaves the bed, and N is the random variable giving the number of ejected grains. The fraction $p_r(\boldsymbol{v}^i)$ of reptating grains is given by $p_r(\boldsymbol{v}^i) = 1 - p_s(\boldsymbol{v}^i)$. By integrating over all possible impact velocities, we find that the fraction of grains in transport that are saltating is

$$\pi_s = \int p_s\big(I(\boldsymbol{v}^0)\big)\, q_+\big(I(\boldsymbol{v}^0)\big)\, f(\boldsymbol{v}^0)\, d\boldsymbol{v}^0/\pi_+, \tag{4.2}$$

where

$$f(\boldsymbol{v}) = \pi_s f_s(\boldsymbol{v}) + (1 - \pi_s)\, f_r(\boldsymbol{v}), \tag{4.3}$$

and where

$$\pi_+ = \int q_+\big(I(\boldsymbol{v}^0)\big)\, f(\boldsymbol{v}^0)\, d\boldsymbol{v}^0 \tag{4.4}$$

is the probability that at least one grain leaves the bed after an impact event. The function I, defined in Section 3, relates the lift-off velocity $\boldsymbol{v}^0$ of a trajectory to its return velocity. Note that f is the density function of the launch velocity vector irrespective of whether the grain is rebounding or ejected.

We are now in a position to give the general equilibrium equations for f_s and f_r. The equation for f_s involves the conditional density function $g_s(v|\boldsymbol{v}^i)$ of the rebound velocity given the impact velocity $\boldsymbol{v}^i$ and given that the impinging grain rebounds. The equation is

$$f_s(\boldsymbol{v}) = \int g_s\big(\boldsymbol{v}|I(\boldsymbol{v}^0)\big)\, p_s\big(I(\boldsymbol{v}^0)\big)\, q_+\big(I(\boldsymbol{v}^0)\big)\, f(\boldsymbol{v}^0)\, d\boldsymbol{v}^0/(\pi_s\pi_+). \tag{4.5}$$

To give the equation for f_r we need the conditional density function $g_r(\boldsymbol{v}|\boldsymbol{v}^i)$ of the ejection velocity given the impact velocity of the impinging grain and given that at least one grain is dislodged.

$$f_r(\boldsymbol{v}) = \int g_r\big(\boldsymbol{v}|I(\boldsymbol{v}^0)\big)\, p_r\big(I(\boldsymbol{v}^0)\big)\, q_+\big(I(\boldsymbol{v}^0)\big)\, f(\boldsymbol{v}^0)\, d\boldsymbol{v}^0 \Big/ \big((1 - \pi_s)\, \pi_+\big). \tag{4.6}$$

The equilibrium launch velocity distribution can for a given splash law be found as the solution to (4.2)—(4.6). In general, there is no guarantee that a unique solution exists, but if the splash law is sufficiently realistic, it is to be expected that a solution can be found by iterating (4.2)—(4.6). Note that through the function I this problem is in general coupled with that of finding the flux ϕ.

Now we seek a realistic splash law with a structure that implies simplification of the equations (4.5) and (4.6). In particular, we aim at equations involving only the vertical component of the take-off velocity vector.

Anderson and Haff's [3] Fig. 2 shows that the mean take-off angle for rebounding grains does not depend much on the impact velocity; see also [4]. This indicates that the horizontal component of the launch velocity v_1 might be conditionally independent of the impact velocity given the vertical component v_2. Mathematically this can be expressed as $g_s(\boldsymbol{v}|\boldsymbol{v}^i)$ $= g_s^{(1)}(v_1|v_2)\, g_s^{(2)}(v_2|\boldsymbol{v}^i)$, which by (4.5) implies that $f_s(\boldsymbol{v}) = g_s^{(1)}(v_1|v_2)\, f_s^{(2)}(v_2)$ for some density function $f_s^{(2)}$. A similar assumption for the ejected grains leaves us with only $f_s^{(2)}$ and $f_r^{(2)}$ to determine. This considerably simplifies solution of the equations above.

Following Werner [17], who found that the conditional density $g_r(\boldsymbol{v}|\boldsymbol{v}^i)$ of the ejection velocity given the impact velocity $\boldsymbol{v}^i$ is nearly independent of $\boldsymbol{v}^i$, we will assume that g_r does not depend on $\boldsymbol{v}^i$. This implies that the launch velocity density for reptators is $f_r(\boldsymbol{v})$ $= g_r(\boldsymbol{v})$.

Werner's [17] distribution of the ejection velocity is indicated in his Fig. 3. Apparently Werner assumes that for reptating grains the two components v_1^0 and v_2^0 are independent, that v_1^0 is normally distributed with mean zero, and that v_2^0 is exponentially distributed. We assume the same, but as the parameter values are not available in [17], we use, as far as possible, values taken from [3]. Thus we arrive at the model $f_r(v_1^0, v_2^0) = f_r^{(1)}(v_1^0)\, f_r^{(2)}(v_2^0)$, where

$$f_r^{(1)}(v_1^0) = \frac{1}{\sqrt{450\pi}} \exp\left[-(v_1^0 - 18)^2/450\right] \tag{4.7}$$

and

$$f_r^{(2)}(v_2^0) = \exp\left(-v_2^0/30\right)/30, \tag{4.8}$$

i.e. v_2^0 has mean 30 cm/s and v_1^0 has mean 18 cm/s and standard deviation 15 cm/s. The standard deviation was read off Werner's Fig. 3. These values are in good agreement with [19].

A final simplification follows if we assume that $g_s^{(2)}(v_2|\boldsymbol{v}^i)$, $p_s(\boldsymbol{v}^i)$ and $q_+(\boldsymbol{v}^i)$ depend only on the vertical component of the impact velocity. For a saltating impinging grain this is not a grossly unrealistic assumption since at equilibrium the impact angle of saltating grains is known not to vary much. When the impinging grain is reptating this assumption is less realistic. If we use the trajectory model of Section 2, the vertical component of the impact velocity depends neither on the horizontal component of the launch velocity nor on the wind speed. Thus the problem of determining $f_s^{(2)}$ is decoupled from that of finding the equilibrium value of ϕ which was treated in Section 3. In particular, this means that f_s is independent of the friction speed. The following equilibrium equation for $f_s^{(2)}$ is obtained by inserting (4.1), (4.3) and (4.8) in (4.5).

$$\pi_+ f_s^{(2)}(v_2)$$

$$= \int_0^\infty g_s\big(v_2|I(v_2{}^0)\big)\, p\big(R|I(v_2{}^0)\big)\, E\big((1+N)^{-1}|I(v_2{}^0)\big)\, f_s^{(2)}(v_2{}^0)\, dv_2{}^0$$

$$+ (1-\pi_s)\,\pi_s^{-1} \int_0^\infty g_s\big(v_2|I(v_2{}^0)\big)\, p\big(R|I(v_2{}^0)\big)\, E\big((1+N)^{-1}|I(v_2{}^0)\big)\, e^{-v_2{}^0/30}\, dv_2{}^0/30. \qquad (4.9)$$

To proceed we need expressions for $p(R|v_2{}^i)$ and $E\big((1+N)^{-1}|v_2{}^i\big)$. Until very recently, there was no information in the literature about the form of the probability distribution of the number N of dislodged grains; see however [4]. Only the mean value of N was studied. We assume here that, given the vertical component $v_2{}^i$ of the impact velocity, N follows a Poisson distribution with mean value $n(v_2{}^i)$, i.e.

$$p(k|v_2{}^i) = n(v_2{}^i)^k \exp\big(-n(v_2{}^i)\big)/k!. \qquad (4.10)$$

From (4.10) it follows that

$$E\big((1+N)^{-1}|v_2{}^i\big) = \big[1 - \exp\big(-n(v_2{}^i)\big)\big]/n(v_2{}^i). \qquad (4.11)$$

For $n(v_2{}^i)$ we use (3.9) modified to depend on $v_2{}^i$, still under the assumption that all grains impact at an angle of $12°$, i.e.

$$n(v_2{}^i) = 0.029 v_2{}^i. \qquad (4.12)$$

However, when the impinging grain is a reptator, we set $n(v_2{}^i) = 0$ to accord with the earlier assumption that reptating grains do not eject other grains on impact.

As mentioned earlier $p(R|v_2{}^i)$ has so far not been treated explicitly in the saltation literature. In [9] it was found that steel spheres shot into a loose bed of similar particles at high speeds rebounded with probability 0.94. Based on this we take the following simple form for $p(R|v_2{}^i)$

$$p(R|v_2{}^i) = \begin{cases} 0.94 & \text{if} \quad v_2{}^i > v_{cr} \\ 0 & \text{if} \quad v_2{}^i \le v_{cr} \end{cases} \qquad (4.13)$$

where the value of v_{cr} must be determined. Willetts and Rice [19] found that grains of medium size dislodge stationary grains with probability 0.7 and that coarse grains do it with probability 0.8. Based on this we guess that a reasonable value for p_s is around 0.8. With $\bar{n} = 1.75$ we find from (3.5) that p_r should be about 0.1. Values of this magnitude are obtained when p_r is calculated by (3.7), (4.8) and (4.13) provided v_{cr} is not far from 50 cm/s.

We shall not here attempt to determine the probability distribution of $f_s^{(2)}$ entirely. Too many aspects of the splash law are still unknown and require informed guess-work. We would therefore not be confident about the finer details of a probability distribution determined by (4.9). Instead we seek a good candidate for $f_s^{(2)}$ in the class of gamma distributions, i.e. we assume that

$$f_s^{(2)}(v) = v^{\alpha-1}\, e^{-\beta v}\beta^\alpha/\Gamma(\alpha) \qquad (4.14)$$

for some $\alpha, \beta > 0$. Here $\Gamma(\alpha)$ denotes the gamma function. The mean value of (4.14) is α/β.

We seek values of α, β and v_{cr} such that the following three equations are satisfied. First, we require that the mean values of the two sides of (4.9) coincide, i.e.

$$\alpha \pi_+ / \beta = \int\limits_0^\infty E_s\big(v_2 | I(v_2{}^0)\big)\, p\big(R | I(v_2{}^0)\big)\, E\big((1 + N)^{-1} | I(v_2{}^0)\big)\, f_s{}^{(2)}(v_2{}^0)\, dv_2{}^0$$

$$+ (1 - \pi_s)\, \pi_s{}^{-1} \int\limits_0^\infty E_s\big(v_2 | I(v_2{}^0)\big)\, p\big(R | I(v_2{}^0)\big)\, e^{-v_2{}^0/30}\, dv_2{}^0/30. \tag{4.15}$$

For π_s we use the value 0.36 found from (3.4) with $\bar{n} = 1.75$ as in Section 3. The mean value $E_s(v_2 | v_2{}^i)$ of the vertical component of the rebound velocity given the vertical component of the impact speed $v_2{}^i$ can be found by simple manipulations of the splash law in [3]:

$$E_s(v_2 | v_2{}^i) = 1.42 v_2{}^i. \tag{4.16}$$

Here we have used that all grains impact at $12°$ and that the average rebound angle is $30°$ irrespective of $v_2{}^i$. The mean rebound angle of $30°$ is the average of results in [3] and [19]. The value of π_+ can be found by integrating over v_2 in equation (4.9). Another possibility is to calculate π_+ from (4.4). Our second requirement is that these two ways of calculating π_+ give the same result. The final equation is the stability condition (3.5) with p_s and p_r calculated by (3.6) and (3.7) and with $\bar{n} = 1.75$.

With I given by (2.10), the solution of our equations is $\alpha = 13$, $\beta = 0.0985$ and $v_{cr} = 60$ cm/s. The mean value of the vertical component of the take-off velocity $v_2{}^0$ for saltating grains is thus found to be 132 cm/s. From (3.6) and (3.7) we find that $p_s = 0.89$ and $p_r = 0.07$, while (4.4) gives $\pi_+ = 0.4$. Using (4.12) we find that $\bar{n} = 2.9$, which is larger than the value $\bar{n} = 1.75$ which we have used in our theory. This discrepancy indicates that the assumptions made about the splash law are not quite consistent. With the present knowledge about the splash law there is no point in trying to correct this deficiency. When $v_2{}^0$ is averaged over reptating as well as saltating grains, we find $E(v_2{}^0) = 67$ cm/s, which compare well with findings in [18], [10], [7] and [15]. By (2.11) we find that the mean jump height for a saltating grains is 6.5 cm. The mean jump height for a grain, irrespective of its transport mode, is 2.6 cm.

In order to calculate the mean values in (3.10), (3.13) and (3.21) we need the conditional expectation of $v_1{}^0$ given $v_2{}^0$ for rebounding grains. If we suppose that $E_s(v_1{}^0 | v_2{}^0)$ is proportional to $v_2{}^0$, it follows from the assumed mean take-off angle of $30°$ that the constant of proportionality is $\cot 30° = 1.73$, i.e.

$$E_s(v_1{}^0 | v_2{}^0) = 1.73\, v_2{}^0. \tag{4.17}$$

For reptating grains we have assumed that the distribution of $v_1{}^0$ is independent of $v_2{}^0$.

5 Refinements

The self-limiting mechanism was rather crudely modelled in Section 3. An obvious refinement would be to use (2.13) for calculating $\bar{n}$ rather than (2.15). Then $\bar{U}$ would vary with U_*, and with a non-constant $\bar{U}$ it would not be reasonable to set $\bar{n}$ constantly equal to 1.75 as was done in Section 3. However, to get expressions for how p_s and p_r depend on the wind profile, we need to know the conditional probability of rebound $p(R | \boldsymbol{v}^i)$, cf. (3.6) and (3.7). If the tentative form (4.13) of $p(R | \boldsymbol{v}^i)$, modified to depend on $v_1{}^i$ rather than on $v_2{}^i$, turns out to be reasonable, it would be relatively easy to calculate $\bar{U}$ and ϕ from (3.5) and (3.17) for a given value of U_*, even if (2.13) is used for calculating $v_1{}^i$. However, simple explicit formulae would not ensue. Note also that to do this, one would have to specify the conditional density $g_s^{(1)}(v_1|v_2^0)$ completely and not just the mean value as was done in Section 4.

The only self-limiting mechanism discussed so far is the retardation of the wind by the cloud of grains in transit. Another possible mechanism, which might be active at high wind speeds, is that as the number of grains in motion becomes large, the probability of rebound decreases so that in turn the number of saltating grains is reduced. This alternative self-limiting mechanism has some times been referred to as the 'soft bed' hypothesis. It is not to be expected that the mechanical properties of the bed itself changes much. A simple calculation, based on findings in [7], shows that even at high transport rates most of the bed is at rest. However, as the number of moving grains becomes large, there will be a relatively large concentration of low-energy ejecta close to the bed and the probability that an impinging grain loses some of its energy by hitting such a grain is not negligible at high transport rates. This mechanism is closely connected to the 'choking' of the transport process discussed in [21]. The concentration of reptating grains close to the bed is proportional to the flux ϕ. Therefore, the importance of the 'soft-bed' in limiting the transport rate could be evaluated by investigating empirically whether $p(R|\boldsymbol{v}^i)$ depends on ϕ. If it does, it would be relatively easy to build this effect into the refined theory discussed above.

Acknowledgement

I am grateful to H. Loft Nielsen for helpful comments and preparation of Fig. 1.

References

[1] Anderson, R. S.: Sediment transport by wind: saltation, suspension, erosion and ripples. Ph.D. Thesis, University of Washington 1986.

[2] Anderson, R. S.: Eolian sediment transport as a stochastic process: the effect of a fluctuating wind on particle trajectories. J. Geol. **95**, 497—512 (1987).

[3] Anderson, R. S., Haff, P. K.: Simulation of eolian saltation. Science **241**, 820—823 (1988).

[4] Anderson, R. S., Haff, P. K.: Wind modification and bed response during saltation of sand in air (this volume).

[5] Anderson, R. S., Sørensen, M., Willetts, B. B.: A review of recent progress in our understanding of aeolian sediment transport (this volume).

[6] Bagnold, R. A.: The physics of blown sand and desert dunes. London: Methuen 1941.

[7] Jensen, J. L., Sørensen, M.: Estimation of some aeolian saltation transport parameters: a reanalysis of Williams' data. Sedimentology **33**, 547—558 (1986).

[8] McEwan, I. K., Willetts, B. B.: Numerical model of the saltation cloud (this volume).

[9] Mitha, S., Tran, M. Q., Werner, B. T., Haff, P. K.: The grain-bed impact process in aeolian saltation. Acta Mech. **63**, 267—278 (1986).

[10] Nalpanis, P.: Saltating and suspended particles over flat and sloping surfaces. II. Experiments and numerical simulation. In: Barndorff-Nielsen, O. E. et al. (eds.): International Workshop on the Physics of Blown Sand, vol. 1, pp. 37—66 (1985).

[11] Owen, P. R.: Saltation of uniform grains in air. J. Fluid Mech. **20**, 225—242 (1964).

[12] Owen, P. R.: The physics of sand movement. Lecture Notes, Workshop on Physics of Desertification, Trieste 1980.

[13] Rasmussen, K. R., Mikkelsen, H. E.: Wind tunnel observations of aeolian transport rates (this volume).

[14] Rasmussen, K. R., Sørensen, M., Willetts, B. B.: Measurement of saltation and wind strength on beaches. In: Barndorff-Nielsen, O. E. et al. (eds.): International Workshop on the Physics of Blown Sand, vol. 2, pp. 301—325 (1985).

[15] Sørensen, M.: Estimation of some aeolian saltation transport parameters from transport rate profiles. In: Barndorff-Nielsen, O. E. et al. (eds.): International Workshop on the Physics of Blown Sand, vol. 1, pp. 141—190 (1985).

[16] Ungar, J. E., Haff, P. K.: Steady state saltation in air. Sedimentology **34**, 289—299 (1987).

[17] Werner, B. T.: A steady-state model of wind-blown sand transport. J. Geol. **98**, 1—17 (1990).

[18] White, B. R., Schulz, J. C.: Magnus effect in saltation. J. Fluid Mech. **81**, 497—512 (1977).

[19] Willetts, B. B., Rice, M. A.: Collision in aeolian transport: the saltation/creep link. In: Nickling, W. (ed.): Aeolian geomorphology. Boston: Allen and Unwin, pp. 1—17 (1986).

[20] Willetts, B. B., Rice, M. A.: Particle dislodgement from a flat sand bed by wind. Earth Surf. Proc. and Land Forms **13**, 717—728 (1988).

[21] Williams, S. H.: A comparative planetological study of particle speed and concentration during aeolian saltation. Ph. D. Thesis, Arizona State University 1987.

Author's address: M. Sørensen, Department of Theoretical Statistics, Institute of Mathematics, University of Aarhus, Ny Munkegade, DK-8000 Aarhus C, Denmark

Acta Mechanica (1991) [Suppl] 1: 83—96

Saltation layers, vegetation canopies and roughness lengths

M. R. Raupach, Canberra, Australia

Summary. This paper argues that vegetation canopies offer an analogue for saltation layers in that both flows have similar momentum sink distributions (from the fluid). Therefore, information on turbulence and momentum transfer in vegetation canopies provides guidance for turbulence in saltation layers. Working from the simplest gradient-diffusion theory for momentum transfer known to be useful in vegetation canopies, the paper explores a first-order theory for fluid momentum transfer in a saltation layer. This leads to an analytic expression for the overall momentum absorption capacity or roughness lenth z_{0S} of the saltation layer, different from previous suggestions and in agreement with field observations of saltation.

1 Introduction

The difficulty of appropriate measurements has meant that data on mean velocity profiles in saltation layers are very scarce, while turbulence data are practically nonexistent. Therefore, in trying to understand how both fluid turbulence and particle motions interact to transfer momentum and mass in saltation layers, it is useful to examine analogues for saltation layers in which the momentum sink (from the fluid) is similar. This paper first argues that vegetation canopies offer such an analogue, so that the copious body of information on turbulence and momentum transfer in vegetation canopies provides some guidance for turbulence in saltation layers. Then, building from the simplest momentum-transfer theory known to be useful in vegetation canopies, the paper examines fluid momentum transfer in a saltation layer, in the light of the complementarity between fluid and particle momentum fluxes which is necessary to conserve momentum through the layer. This leads to a simple gradient-diffusion hypothesis about momentum transfer through the air in a saltation layer, which indicates how the saltation trajectories and the airflow modify one another.

The momentum-transfer hypothesis leads to an analytic estimate of the overall momentum-absorbing capacity or roughness length z_{0S} of the saltation layer. The derived expression extends the relation suggested by Charnock [1], Owen [2] and Chamberlain [3], $z_{0S} = Cu_*^2/2g$ (where u_* is the friction velocity, g the gravitational acceleration and $C \approx 0.02$ is an empirical constant), by rationally approaching the limit of small u_*. The expression turns out to give results in good agreement with recent field experiments by Rasmussen et al. [22] and with the latest numerical models of the saltation process (R. S. Anderson, personal communication, 1990) but somewhat in contradiction with the available data from wind tunnels.

2 Particle and fluid momentum transfer

In a saltation layer, the total shear stress or momentum flux density (τ) consists of contributions from both the particle motion (τ_p) and the turbulence (τ_a). In the absence of advection or streamwise pressure gradients, momentum conservation requires that the total momentum flux $\tau(z)$ is constant both above and through the saltation layer, which defines the friction velocity $u_* = (\tau/\varrho)^{1/2}$ (ϱ being the air density). It follows that τ_p and τ_a vary in a complementary way:

$$\tau = \varrho u_*^2 = \tau_p + \tau_a \tag{1}$$

The particle momentum flux density τ_p is defined by the particle trajectories, specified by the streamwise and vertical particle velocities $U(t)$, $W(t)$ and displacements $X(t)$, $Z(t)$, with t being travel time. Basic momentum balance considerations show that the particle momentum flux density $\tau_p(z)$ at height z above the surface is given by:

$$\tau_p(z) = \sum_{i=1}^{n} m_i\big(U_{\downarrow i}(z) - U_{\uparrow i}(z)\big) \tag{2}$$

where n is the vertical flux density of particles (either up or down), m is particle mass, $U_{\uparrow}(z)$ and $U_{\downarrow}(z)$ are streamwise particle velocities at height $z = Z(t)$ during the upward and downward parts of particle trajectories, respectively, and the subscript i enumerates particles crossing unit area in unit time. At the surface ($z = 0$), n is called the ejection rate or the dislodgement rate.

In the highly idealized case of identical particles undergoing identical trajectories, as assumed in the saltation models of Owen [2] and Ungar and Haff [4], Eq. (2) can be simplified to

$$\tau_p(z) = nm\big(U_{\uparrow}(z) - U_{\downarrow}(z)\big) \tag{3}$$

However, it is essential in practice to allow variation among the particle trajectories. If the trajectories differ only in initial vertical particle velocity W_0, then Eq. (2) becomes

$$\tau_p(z) = mn \int_0^{\infty} P(W_0)\big(U_{\downarrow}(z; W_0) - U_{\uparrow}(z; W_0)\big)\, dW_0 \tag{4}$$

where $P(W_0)$ is the probability density function of W_0, which has been argued by Anderson and Hallet [5] to be close to the exponential distribution

$$P(W_0) = \frac{1}{\langle W_0 \rangle} \exp\left(\frac{-W_0}{\langle W_0 \rangle}\right) \tag{5a}$$

where $\langle W_0 \rangle$ is the mean initial vertical velocity of the particles, which can be taken to scale as

$$\langle W_0 \rangle = \alpha u_*. \tag{5b}$$

Anderson and Hallet [5] suggested $\alpha = 0.63$. Sørensen [24] also found an approximately exponential distribution for W_0.

Figure 1 shows sketches of the height variation of τ_p and τ_a produced in saltation layers with identical and nonidentical particle trajectories [Eqs. (3) and (4), respectively]. When the

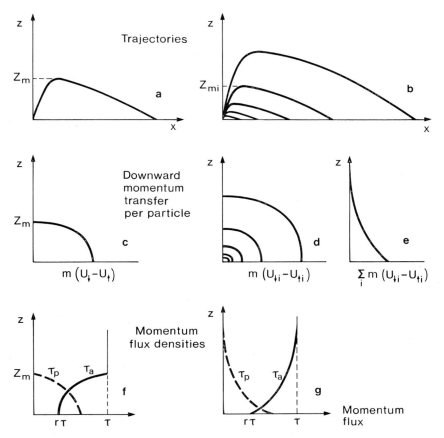

Fig. 1. Saltation-layer characteristics for cases of identical trajectories (left-hand column) and nonidentical trajectories (right-hand column). **a** and **b**, typical saltation trajectories; **c** and **d**, profiles against height z of downward momentum transfer per particle; **e** sum of downward particle momentum transfers over all particles, for nonidentical trajectory case; **f** and **g**, profiles against z of total momentum flux density τ, particle-borne momentum flux density τ_p and airborne momentum flux density τ_a

trajectories are identical, the strong exchange of momentum near the top of the single prototype trajectory (at $z = Z_m$) causes $\tau_a(z)$ to decrease with z most strongly just below $z = Z_m$. On the other hand, the spread of Z_m values in the nonidentical case causes the strongest decrease of $\tau_a(z)$ with z to occur near the ground. The result is that the profiles of $\tau_a(z)$ have opposite curvatures in the two cases. Measured profiles of τ_p, inferred from wind-tunnel data [24] and data on beach saltation [22], are qualitatively very similar to the non-identical particle case in Fig. 1.

Whatever the precise shapes of the profiles of $\tau_p(z)$ and $\tau_a(z)$ in the saltation layer, the primary effect of the particles on the airflow is to provide a spatially distributed momentum sink, with sink strength or density $d\tau_a/dz = -d\tau_p/dz$ per unit volume of fluid. This sink modifies both the mean wind profile and the turbulence in the saltation layer. To obtain empirical guidance on how the wind field is modified, and how such a flow can be described theoretically, it is useful to look at other turbulent shear flows with spatially distributed momentum sinks; a well-studied example is the flow in a vegetation canopy.

3 The wind field in a vegetation canopy

A large number of experiments on canopy turbulence, both in the field and in wind tunnel models, have by now produced a fairly consistent description of the flow both within and above a canopy, reviewed in [6], [7] and [8]. Drawing from these reviews, which contain a great deal more detail than is mentioned here, the main features of the turbulent flow within and just above a vegetation canopy can be summarized by reference to Fig. 2. This shows measured profiles of air velocity moments and turbulence length scales from seven canopies: two forests, two corn canopies and three wind tunnel models (see [6], [7], [8] for details of the canopies). Here and subsequently, the air velocity vector is (u, v, w), with overbars denoting (time) averages and primes fluctuations therefrom; the coordinate system (x, y, z) has x in the mean wind direction and z vertical, with the ground at $z = 0$. The normalizing length and velocity scales are the canopy height h and the friction velocity u_*.

It is first noteworthy that normalization with u_* and h is fairly successful in collapsing data from canopies in which h varies by a factor of over 400 and u_* by a factor of more than 10. This suggests that u_* is the dominant velocity scale for canopy turbulence, and that h (or a closely related scale) is the dominant length scale.

Next, there is strong shear in the mean wind $\overline{u}(z)$ near the canopy top $z = h$ (Fig. 2a), inducing a pronounced inflection in the mean wind profile near $z = h$. The Reynolds shear stress $\overline{u'w'}(z)$ $(= -\tau_a/\varrho)$ is constant above the canopy (such that $-\overline{u'w'} = u_*^2$) but attenuates rapidly with depth within the canopy (Fig. 2d). The u and w standard deviations, σ_u and σ_w, behave in a qualitatively similar way (Figs. 2b, c), though the details are somewhat different.

Thirdly, the u and w skewnesses Sk_u and Sk_w are both large, with $Sk_u > 0$ and $Sk_w < 0$ within the canopy (Figs. 2e, f), indicating that the strongest turbulent events are gusts: energetic, downward incursions of air into the canopy space from the faster-moving flow above. It can be shown by quadrant analysis of both field data, [9], [10], [11], and laboratory data, reviewed in [8], that the gusts are responsible for most of the momentum transfer, but are also highly intermittent; typically, more than 50% of the momentum transfer occurs in less than 5% of the time.

Fourthly, the dominant canopy eddies (of which gusts form the most energetic and active part) are coherent over streamwise and vertical distances of order h. This is shown in Figs. 2g, h by measurements of the turbulence length scales L_u and L_w (found from the single-point Eulerian u and w time scales with a frozen-turbulence hypothesis): near $z = h$, $L_u \approx h$ and $L_w \approx h/3$.

Finally, spectral data and scale analyses from [12] and [6], not reproduced here, show clearly that the dominant canopy eddies are not associated with element wakes, which are too small and short-lived to transport much momentum or contribute much energy. Instead, the dominant canopy eddies are directly associated with shear on vertical scales of order h. This implies that the canopy turbulence structure is relatively insensitive to the precise geometry of the momentum-absorbing canopy elements, as is found in practice.

To describe satisfactorily the vertical transfer of momentum and mass by canopy turbulence, it is necessary to account for the fact that the vertical length scale of the turbulence is comparable with the vertical scales over which mean gradients change appreciably. This makes questionable a gradient-diffusion transfer description [13], which for momentum reads

$$\tau_a/\varrho = K \, d\overline{u}/dz \tag{6}$$

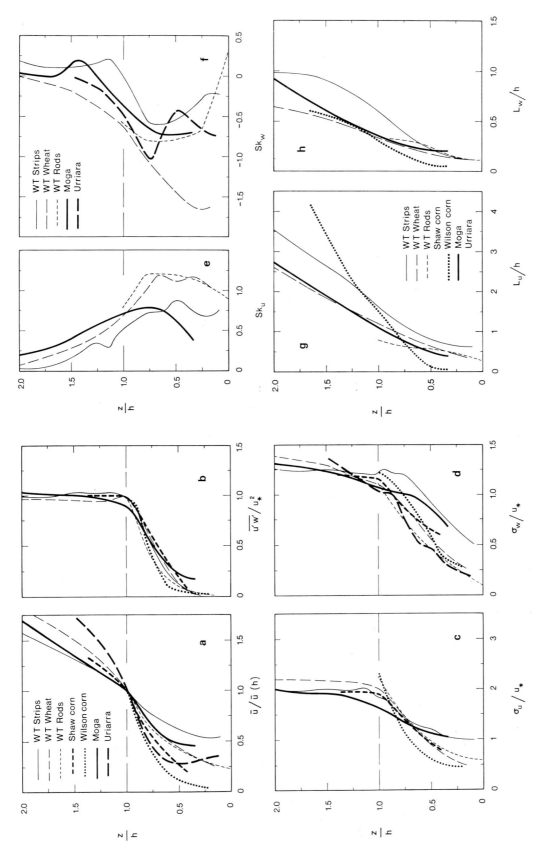

Fig. 2. Profiles of mean air velocity, turbulent velocity moments and turbulence length scales from seven vegetation canopies. See text for definitions of symbols. Canopy details in [6], [7], [8]

where K is an eddy diffusivity. Indeed, countergradient fluxes of scalar entities (heat, water vapour, CO_2) are observed clearly in vegetation canopies under some circumstances [14]. Countergradient momentum fluxes are also likely to exist, even though the observational evidence is not as clear [6]. In either case, countergradient fluxes imply negative (and also highly erratic) values of the eddy diffusivity K.

To improve on the gradient-diffusion transfer description of Eq. (6), two theoretical approaches have been explored: Eulerian higher-order closure theories (which apply for both scalars and momentum) and Lagrangian theories (which apply to scalars only, since the wind field is assumed to be known *a priori*). The basis of each approach is reviewed by Finnigan and Raupach [15]. Eulerian higher-order closure schemes require numerical solution, but the Lagrangian approach can be cast in a sufficiently simple form to offer some analytic insight into the effects of the large length scales (or persistence times) of canopy turbulence.

Raupach [7], [16] used Lagrangian methods to express the concentration of an arbitrary scalar as the sum of a far-field, diffusive contribution, obeying a gradient-diffusion theory, and a near-field, nondiffusive contribution related to the proximity of scalar sources to a point of observation. Two outcomes from this work are relevant here: firstly, the theory shows that the nondiffusive aspects of the transfer are relatively localized in the vertical to layers close to the major canopy sources or sinks, and that the larger-scale features of the concentration field behave diffusively to a good approximation. Secondly, the theory distinguishes two generic types of transfer problem: the "forward" problem of finding a concentration profile from a given profile of scalar source or sink density, and the "inverse" problem of finding the source or sink density from a given concentration profile. Of these two, the "forward" problem is by far the more tractable (or mathematically well-posed) in the sense that the output is less sensitive to small perturbations in the input. In the same way, "forward" solutions are far less sensitive to the precise description of turbulent transfer, compared with "inverse" solutions. A gradient-diffusion assumption like Eq. (6) can often give acceptable approximate solutions to the "forward" problem, even though its solutions to the "inverse" problem are grossly inaccurate.

In the case of momentum transfer, the Lagrangian theory cannot be applied directly. Nevertheless, the same distinction between "forward" and "inverse" problems remains valid. The prediction of the mean velocity profile (momentum concentration profile) $\bar{u}(z)$ in the canopy is a "forward" problem, which suggests that a gradient-diffusion approximation is tenable, provided the eddy diffusivity K is empirically adjusted to account for near-field effects. This conclusion is supported by the fact that solutions of Eq. (6) have long been used to give simple, fairly realistic expressions for $\bar{u}(z)$ in canopies, which are not very sensitive to the precise choice of K [17].

4 The wind field in a saltation layer

According to the arguments in Section 2, the fluid momentum sink density $-\varrho \, \overline{du'w'}/dz = d\tau_a/dz$ is vertically distributed both in a saltation layer and in a vegetation canopy (though the strongest sink occurs near the ground in a saltation layer, and near $z = h$ in a canopy). Furthermore, wake turbulence in a saltation layer should be a small contributor to the overall transport and turbulent energy, just as in a canopy, because the wake turbulence is very small-scale and short-lived in both cases.

These similarities are sufficient to suggest that the main features of turbulence in canopies should also apply in saltation layers. In particular, to the same extent and with the same reservations as in a vegetation canopy, Eq. (6) should offer a reasonable approximate solution to the "forward" problem of finding the mean wind profile $\overline{u}(z)$ in a saltation layer, given a specification of the momentum sink density $-\varrho \, \overline{du'w'}/dz = d\tau_a/dz$. It is necessary to choose the eddy diffusivity K, which may be done by writing $K = w_s L_s$, where w_s is an eddy velocity scale and L_s an eddy length scale. Reasonable choices for a saltation layer are:

$$w_s = (\tau_a/\varrho)^{1/2} = (-\overline{u'w'})^{1/2} \tag{7a}$$

$$L_s = kz \tag{7b}$$

where k is the von Karman constant ($k = 0.40$). These choices make K converge properly to its surface-layer value ku_*z above the saltation layer, and are broadly consistent with canopy eddy velocity and length scale data. Equation (6) now becomes:

$$\frac{d\overline{u}}{dz} = \frac{1}{kz}\left(\frac{\tau_a}{\varrho}\right)^{1/2} = \frac{1}{kz}\left(\frac{\tau - \tau_p}{\varrho}\right)^{1/2} \tag{8}$$

where τ_p is specified from the particle trajectories by Eq. (2). Equation (8) makes explicit the particle-airflow interaction, by showing how the particle trajectories modify the wind profile $\overline{u}(z)$. This equation was first suggested by Anderson [23] and has been used in recent numerical models of saltation [18].

5 Saltation models

At this point it is useful to consider the above equations in the context of a comprehensive model of the saltation process. Anderson and Haff [18] have identified four components, or subprocesses, in such a model: (a) aerodynamic liftoff of particles from the bed; (b) particle trajectories; (c) particle impact or splash, including bounce of the impacting particle and the ejection of other particles; and (d) particle-airflow interaction, or modification of the wind field by the grains. The interaction between these four subprocesses provides a self-limiting mechanism, which determines the eventual equilibrium saltation state of a mobile surface subjected to a given wind field (defined, say, by the total shear stress τ). Most of the foregoing has concerned (d).

To specify trajectories, (b), most saltation models (e.g. [2], [5], [18]) consider the grains as spherical particles moving under the influence of only fluid drag and gravity, so that the trajectories are given by:

$$\frac{dU}{dt} = -\frac{3\varrho}{4\varrho_p d} C_D(\text{Re}) \, q\big(U - \overline{u}(z)\big) \tag{9}$$

$$\frac{dW}{dt} = -\frac{3\varrho}{4\varrho_p d} C_D(\text{Re}) \, qW - g \tag{10}$$

where ϱ_p is the particle density, d the particle diameter, g the acceleration due to gravity, $q = \big((U - \overline{u})^2 + W^2\big)^{1/2}$ the relative particle velocity, $\text{Re} = qd/v$ the particle Reynolds number, v the kinematic viscosity of air and $C_D(\text{Re})$ the drag coefficient for a sphere. Lift and Magnus (spin) forces on the particle are ignored, following Anderson and Hallet [5], who showed these forces to be fairly small and tending to oppose one another. Also, Eqs. (9)

and (10) include only the mean fluid velocity and ignore the turbulence, a common approximation in saltation (but not suspension) studies.

Aerodynamic liftoff and splash, (a) and (c), have been modelled by various workers in a variety of ways. Anderson and Haff [18] modelled splash with a "splash function" which determines probabilistically the outcome of a particle impact on the bed, using a two-dimensional model of the grain mechanics within the bed. The self-limiting mechanism in their model is provided essentially by the competition between impact-induced particle splash, (c), which (in a fixed wind field) tends to increase the number of particles in motion as time proceeds, and particle-air interactions, (d), which reduce the wind speed near the bed until equilibrium is reached. In this view, aerodynamic liftoff plays little role in fully developed, equilibrium saltation, almost all particle motion being initiated by impacts. For the cases they modelled, predicted values of $\tau_a(0)$ at equilibrium were slightly less than the fluid threshold shear stress $\varrho u_{*t\text{(init)}}^2$, where $u_{*t\text{(init)}}$ is the threshold friction velocity required to initiate particle motion over a surface at rest.

Much earlier, in 1964, Owen [2] had advanced a different idea on the self-limiting mechanism, in which $\tau_a(0)$ plays a central role. He made the hypothesis that, in equilibrium saltation, the air shear stress at the ground is just large enough to "ensure that the surface grains are in a mobile state", so that $\tau_a(0) = \varrho u_{*t}^2$, where u_{*t} is the threshold friction velocity necessary to keep an already mobilized surface in motion (slightly less than $u_{*t\text{(init)}}$). This hypothesis implies that fluid (rather than impact) forces are dominant in maintaining the saltation (see [2], p. 228). It follows that equilibrium state satisfies

$$\tau_p(0) = \tau(0) - \tau_a(0) = \varrho u_*^2(1 - u_{*t}^2/u_*^2) \tag{11}$$

Combining Eqs. (11) and (4) gives

$$\tau_p(z) = \varrho u_*^2 \left(1 - \frac{u_{*t}^2}{u_*^2}\right) \frac{\displaystyle\int_0^\infty P(W_0)\left(U_\downarrow(z; W_0) - U_\uparrow(z; W_0)\right) dW_0}{\displaystyle\int_0^\infty P(W_0)\left(U_\downarrow(0; W_0) - U_\uparrow(0; W_0)\right) dW_0} \tag{12}$$

which yields $\tau_p(z)$ and thence $\tau_a(z) = \varrho u_*^2 - \tau_p(z)$ from computed particle trajectories and given values for u_* and u_{*t}. The ejection rate n has been eliminated using Eq. (11), but can, of course, be calculated *post facto*. In this model, neither aerodynamic liftoff (a) nor splash (c) are modelled directly; instead, the equilibrium state is determined entirely by τ_a.

Besides different hypotheses about the significance of τ_a, a further important difference between Anderson and Haff's and Owen's models is that the former can predict the process by which equilibrium is approached, while the latter predicts only the equilibrium state. Restriction to equilibrium saltation makes the Owen model relatively simple; in fact, its original form was analytic, with linearization of the trajectory equations, Eqs. (9) and (10), and a different, simplified form of Eq. (8) ($\bar{u}(z) = $ constant). These assumptions are easily removed by admitting a numerical component in the model[1].

The quantitative difference between the hypotheses of Anderson and Haff (that impacts are critical in determining the equilibrium state) and Owen (that fluid forces and τ_a are

---

[1] I have constructed a simple numerical version of the Owen model which removes the need to linearize Eqs. (9) and (10). From specified u_* and u_{*t} values, particle trajectories are computed using Eqs. (9) and (10), with an unperturbed $\bar{u}(z)$ (with $\tau_p(z) = 0$); then $\tau_p(z)$ is found from Eq. (2), giving a modified $\bar{u}(z)$ from Eq. (8). The process is iterated to convergence, typically needing a modest 5 to 7 iterations.

dominant) is actually less than might appear at first. In the equilibrium states predicted by both models, the air shear stress at the ground is similar: it is ϱu_{*t}^2 in Owen's model, and is slightly less than $\varrho u_{*t(\text{init})}^2$ in Anderson and Haff's model. This is consistent with Bagnold's classical distinction [19] between the "fluid" threshold friction velocity $u_{*t(\text{init})}$ and the "impact" threshold friction velocity $u_{*t} \approx 0.8 u_{*t(\text{init})}$ required to maintain pre existing saltation.

6 Overall roughness length of the saltation layer

One important consequence of the particle-airflow interaction in saltation is the partition of the total momentum flux density into particle-borne and airborne components, τ_p and τ_a. A second is the behaviour of the roughness length z_{0S} of a saltating surface: as wind speed increases, z_{0S} becomes progressively greater than the roughness length z_0 of the same surface without saltation. These two effects are related: the total momentum flux density τ to the surface, and therefore z_{0S}, both increase as the particle-borne component τ_p increases. Since z_{0S} is a readily observable quantity, whereas the partition between τ_p and τ_a is much harder to measure, it is useful to know exactly how z_{0S} is related to the partition between τ_p and τ_a, so that interferences about the internal dynamics of the saltation layer can be drawn from measurements of z_{0S}. The purpose of this final section is to establish and analyse such a relationship.

Rather than using a numerical model (of either of the above forms), it is simpler and more instructive to use an analytic expression for $\tau_a(z)$, satisfying the same global constraints as the numerical models; the results can be checked against the models *post facto*. The available global constraints are

(a) Eq. (1), $\tau_p(z) + \tau_a(z) = \tau = \varrho u_*^2$.

(b) $\tau_p \to 0$ and $\tau_a \to \tau$ as $z \to \infty$.

(c) τ_p and τ_a are both monotonic with height and have shapes compatible with Fig. 1g.

(d) The characteristic height H_S of the τ_p profile is of the order of the mean particle jump height $\langle Z_m \rangle$, so that $H_S = \beta \langle Z_m \rangle$, where β is an $O(1)$ constant. In turn, $\langle Z_m \rangle$ may be represented by the ballistic approximation $\langle W_0 \rangle^2/(2g)$, which has been shown to give a reasonable (albeit slightly low) estimate of trajectory heights [5]. The mean initial particle velocity $\langle W_0 \rangle$ can be represented by Eq. (5b), giving

$$H_S = \beta \alpha^2 u_*^2/(2g) \tag{13}$$

(e) At $z = 0$, $\tau_a(0) = \varrho u_{*t}^2$. This is the Owen self-limiting hypothesis [2], but (as indicated above) the difference in practice between this and the Anderson and Haff [18] model result is small. The essential point is that $\tau_a(0)$ in a saltation layer takes an externally precribed value, here assumed to be ϱu_{*t}^2.

A suitable expression for $\tau_a(z)$, which satisfies all these constraints and is analytically convenient, is

$$\left(\frac{\tau_a(z)}{\varrho u_*^2}\right)^{1/2} = \frac{w_s}{u_*} = 1 - \left(1 - \sqrt{r}\right) e^{-z/H_S} \tag{14}$$

where w_s is the eddy velocity scale defined in Eq. (7a), and the ratio $r = \tau_a(0)/\tau$; r is a descriptor of the partition between particle-borne and airborne stresses at the surface. Using constraint (e) above, it follows that $r = u_{*t}^2/u_*^2$ for $u_* \geq u_{*t}$, and $r = 1$ for $u_* < u_{*t}$.

Substituting Eq. (14) into Eq. (8), one obtains the mean velocity shear

$$\frac{d\overline{u}}{dz} = \frac{u_*}{k}\left[\frac{1}{z} - \left(1 - \sqrt{r}\right)\frac{\exp\left(-z/H_S\right)}{z}\right]$$

which integrates to

$$\overline{u}(z) = \frac{u_*}{k}\left[\ln\left(\frac{z}{z_0}\right) - \left(1 - \sqrt{r}\right)\left(E_1\left(\frac{z_0}{H_S}\right) - E_1\left(\frac{z}{H_S}\right)\right)\right] \tag{15}$$

where z_0 is the roughness length of the underlying surface $\left(\text{such that } \overline{u}(z_0) = 0\right)$ and $E_1(x)$ is the exponential integral defined by

$$E_1(x) = \int\limits_x^\infty \frac{e^{-t}}{t}\, dt$$

and with the properties $E_1(x) \to -\gamma - \ln(x)$ as $x \to 0$ (where $\gamma = 0.577216$ is Euler's constant) and $E_1(x) \sim \exp\left(-x\right)/x \to 0$ as $x \to \infty$.

Restricting consideration to the wind profile well above the saltation layer $(z \gg H_S)$ the second E_1 term in Eq. (15) vanishes. Provided also that $z_0 \ll H_S$, which is certainly the case in well-developed saltation, then the first E_1 term can be replaced by its small-argument expansion, giving

$$\overline{u}(z) = \frac{u_*}{k}\left[\ln\left(\frac{z}{z_0}\right) + \left(1 - \sqrt{r}\right)\left(\gamma + \ln\left(\frac{z_0}{H_S}\right)\right)\right] = \frac{u_*}{k}\ln\left(\frac{z}{z_{0S}}\right) \tag{16}$$

where the second equality is the definition of the saltation-layer roughness length z_{0S}. It follows that

$$\ln\left(z_{0S}\right) = \left(1 - \sqrt{r}\right)\ln\left(H_S\right) + \sqrt{r}\ln\left(z_0\right) - \left(1 - \sqrt{r}\right)\gamma \tag{17}$$

so that, using Eq. (13),

$$z_{0S} = \left(A\,\frac{u_*^2}{2g}\right)^{1 - \sqrt{r}} z_0^{\sqrt{r}} \tag{18}$$

where $A = e^{-\gamma}\beta\alpha^2$ is an $O(1)$ constant $\left(\text{and } \sqrt{r} = u_{*t}/u_*\right)$. This is the required expression for z_{0S} in terms of dynamical properties of the saltation layer. If $\beta = 1$ (a likely value[2]) and $\alpha = 0.63$ (from Eq. (5b)), then $A = 0.22$.

Equation (18) shows that z_{0S} is a weighted geometric mean (with unequal exponents summing to 1) of the undisturbed roughness length z_0 and a length $Au_*^2/(2g)$ proportional to the characteristic height H_S of the saltation layer (or more strictly the height of the τ_p and τ_a profiles). When there is no saltation, $u_* \leq u_{*t}$ and $\sqrt{r} = 1$, so $z_{0S} = z_0$; at very high wind speeds ($u_* \to \infty$), $\sqrt{r} \to 0$ and $z_{0S} \to Au_*^2/(2g)$. The behaviour of z_{0S} with wind speed, for surfaces with $z_0 = 0.01$, 0.1 and 1 mm, is shown in Fig. 3 for two choices of A (0.2 and 0.3), assuming $u_{*t} = 0.2$ m s^{-1}, a typical value for aeolian sand.

From Eq. (18), one obtains

$$C = \frac{z_{0S}}{u_*^2/2g} = A^{1 - \sqrt{r}}\left(\frac{z_0}{u_*^2/2g}\right)^{\sqrt{r}} \tag{19}$$

---

[2] Other estimates of $\alpha_1 = 2g < Z_m > /u_*^2$ are higher than 1; for instance, Nalpanis [20] obtained $\alpha_1 = 2.2$ with photographic techniques. However, photography may be biased towards faster particles [5] and therefore to higher α_1 values

This result may be compared with the well-known suggestion of Owen [2] and others[3] that C is constant, at $C \approx 0.016$. Equation (19) predicts that C is not constant but depends strongly on u_*, since $\sqrt{r} = u_{*t}/u_*$. The variation of C with u_* is shown in Fig. 4, for the same conditions as in Fig. 3.

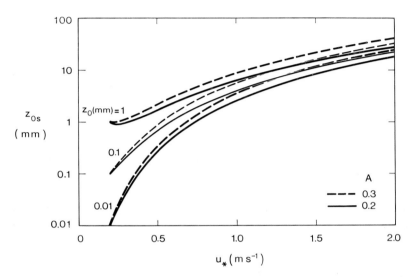

Fig. 3. Prediction of Eq. (18) for z_{0S} over surfaces with undisturbad roughness length $z_0 = 0.01$, 0.1 and 1.0 mm, and with dimensionless constant $A = 0.2$ (solid lines) and 0.3 (broken lines). Parameter values: $u_{*t} = 0.2$ m s^{-1}, $g = 9.8$ m s^{-2}

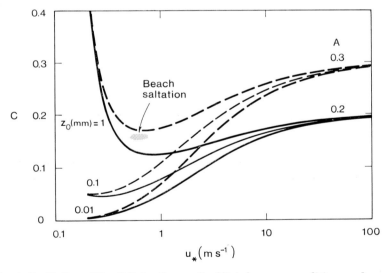

Fig. 4. Prediction of Eq. (19) for $C = z_{0S}/(u_*{}^2/2g)$, for same conditions and parameter values as Fig. 3. Data from measurements of beach saltation [22]

---

[3] The same equation had earlier been suggested by Charnock [1] for the roughness of a sea surface in the presence of flying spray, and was suggested as a general result for mobile (sand, sea, snow) surfaces by Chamberlain [3], though the "constant" C showed considerable variability

Several aspects of Eqs. (18) and (19) are significant.

(a) Eq. (18) properly gives $z_{0S} = z_0$ when $u_* \leq u_{*t}$, in contrast with the equation $z_{0S} = C u_*^2/(2g)$ with constant C, which is unrealistic as $u_* \to 0$ (predicting $z_{0S} \to 0$).

(b) Equations (18) and (19) predict that both C and z_{0S} are independent of the proportionality constant A at low $u_* \left(\sqrt{r} \to 1 \right)$, and independent of the undisturbed roughness length z_0 at high $u_* \left(\sqrt{r} \to 0 \right)$. Therefore, at low u_* values, uncertainty in A is not fully reflected in z_{0S}.

(c) At very high u_* values, $C \to A$, so Eq. (19) appears compatible with the suggestion of constant C. However, the required values of u_* for this limit to be reached are unrealistically large, as shown in Fig. 4 where the plot has been continued to extremely high u_* values in order to show the manner in which $C \to A$. Therefore, this feature of Eq. (19) does not support the suggestion of constant C in practice.

(d) At realistic values of u_*, Eq. (19) predicts an interesting and rather complicated behaviour for C. There is a strong dependence on the undisturbed roughness length z_0, since at threshold, $C = z_0/(u_{*t}^2/2g)$. This implies that for surfaces with moderate or larger z_0 (greater than about 0.3 mm), C decreases rapidly with increasing u_* in the range just above threshold, before increasing again at larger u_* values. Hence, there is a range of u_* values, near the minimum in C, where the change of C with u_* is very small. Since this u_* range coincides quite well with naturally occurring values, this is a possible explanation for observations of nearly constant C in practice.

(e) At sufficiently large z_0 and A values, Eq. (18) predicts that, just above threshold, z_{0S} actually decreases with increasing u_* (see Fig. 3 for the case $z_0 = 1$ mm, $A = 0.3$). This implies that, as u_* increases over these rough surfaces, the developing saltation layer at first shelters the static roughness on the surface before introducing extra drag and roughness of its own. It is not known whether such behaviour is observed in practice.

Finally, Eq. (19) may be compared with observations and other models. Field experiments on beach saltation by Rasmussen and colleagues [22] produced a tightly grouped data set giving $z_{0S} \approx 4$ mm and $C \approx 0.16$ over a rather narrow range of wind speeds (u_* between 0.6 and 0.7 m s^{-1}). These data are shown on Fig. 4. Although an undisturbed z_0 measurement is not available, a value between 0.3 and 1 mm is plausible for a beach surface. Hence, these data appear consistent with the theory, and also suggest that a choice for the constant A in the range 0.2 to 0.3 is reasonable.

Also consistent with the theory are recent model results from the model of Anderson and Haff [18] (R. S. Anderson, personal communication, 1990). Like Eq. (19), the model predictions produce C values between 0.1 and 0.2 in typical conditions. This is perhaps not surprising, since both this theory and the model use Eq. (7) as a basic description of the fluid momentum transfer.

Inconsistent results appear to be produced by observations of z_{0S} in saltation wind tunnels, where C is often observed to be about 0.02. Owen [2] suggested a value of this order from analysis of early wind-tunnel data on saltation; recently, Rasmussen [21] has obtained a similar value from the Aarhus saltation tunnel. A possible explanation for the inconsistency, advanced by Anderson and Haff [18], is that the fetches of existing saltation wind tunnels (10 m or less) are too short to produce equilibrium saltation; the critical feedback of the particles on the airflow, which stabilizes the saltation process in their model, does not develop within the available fetch. Hence, the observed wind profiles are determined by some upstream state rather than by the equilibrium saltation layer.

7 Conclusions

It has been argued that turbulence, and the consequent vertical transfer of streamwise momentum by the fluid, is essentially similar in saltation layers and vegetation canopies. A gradient-diffusion description of fluid momentum transfer, although flawed, is useful in both cases.

Using this, together with the complementarity between fluid and particle momentum transfer in a saltation layer, a relationship has been derived between the roughness length z_{0S} of the saltation layer and the stress partition at the ground surface, expressed by the ratio $r = \tau_a(0)/\tau$. Following Owen [2], and in approximate agreement with the recent model results of Anderson and Haff [18], it has been assumed that $\tau_a(0)$ is equal to the (impact) threshold shear stress ϱu_{*t}^2.

The resulting relationship shows that z_{0S} is equal to the undisturbed roughness length z_0 at low wind speeds (at and below threshold), and approaches $A u_*^2/(2g)$ at extremely high wind speeds, with A a constant around 0.2 to 0.3. The resulting behaviour of the dimensionless quantity $C = z_{0S}/(u_*^2/2g)$ is subtle, with C approximately independent of u_* in typical conditions.

These results agree reasonably with field data suggesting $C \approx 0.16$ for beach saltation, but poorly with wind tunnel data, which suggest $C \approx 0.02$. A possible reason for the discrepancy is that wind tunnel saltation is developed over too short a fetch to reach equilibrium.

References

[1] Charnock, H.: Wind stress on a water surface. Quart. J. Roy. Meteorol. Soc. **81**, 639—640 (1955).

[2] Owen, P. R.: Saltation of uniform grains in air. J. Fluid Mech. **20**, 225—242 (1964).

[3] Chamberlain, A. C.: Roughness length of sea, sand and snow. Boundary-Layer Meteorol. **25**, 405—409 (1983).

[4] Ungar, J., Haff, P. K.: Steady state saltation in air. Sedimentology **34**, 289—299 (1987).

[5] Anderson, R. S., Hallet, B.: Sediment transport by wind: toward a general model. Geol. Soc. Am. Bull. **97**, 523—535 (1986).

[6] Raupach, M. R.: Canopy transport processes. In: Flow and transport in the natural environment: advances and applications. (eds. W. L. Steffen, O. T. Denmead) Springer, Berlin, pp. 95—127 (1988).

[7] Raupach, M. R.: Stand overstorey processes. Phil. Trans. Roy. Soc. London B **324**, 175—190 (1989).

[8] Raupach, M. R., Antonia, R. A., Rajagopalan, S.: Rough-wall turbulent boundary layers. Appl. Mech. Rev. **44**, 1—25 (1991).

[9] Finnigan, J. J.: Turbulence in waving wheat. I. Mean statistics and honami. Boundary-Layer Meteorol. **16**, 181—211 (1979).

[10] Finnigan, J. J.: Turbulence in waving wheat. II. Structure of momentum transfer. Boundary-Layer Meteorol. **16**, 213—236 (1979).

[11] Shaw, R. H., Tavangar, J., Ward, D. P.: Structure of the Reynolds stress in a canopy layer. J. Climate Appl. Meteorol. **22**, 1922—1931 (1983).

[12] Shaw, R. H., Silversides, R. H., Thurtell, G. W.: Some observations of turbulence and turbulent transport within and above plant canopies. Boundary-Layer Meteorol. **5**, 429—449 (1974).

[13] Corrsin, S.: Limitations of gradient transport models in random walks and in turbulence. Adv. Geophys. **18 A**, 25—60 (1974).

[14] Denmead, O. T., Bradley, E. F.: On scalar transport in plant canopies. Irrig. Sci. **8**, 131—149 (1987).

[15] Finnigan, J. J., Raupach, M. R.: Transfer processes in plant canopies in relation to stomatal characteristics. In: Stomatal function. (eds. E. Zeiger, G. D. Farquhar, I. R. Cowan). Stanford University Press, Stanford, CA, USA, pp. 385—429 (1987).

[16] Raupach, M. R.: Applying Lagrangian fluid mechanics to infer scalar source distributions from concentration profiles in plant canopies. Agric. For. Meteorol. **47**, 85—108 (1989).

[17] Thom, A. S.: Momentum, mass and heat exchange of plant communities. In: Vegetation and the atmosphere, vol. 1. (ed. J. L. Monteith), Academic Press, London, pp. 57—109 (1975).

[18] Anderson, R. S., Haff, P. K.: Wind modification and bed response during saltation of sand in air (this volume).

[19] Bagnold, R. A.: The physics of blown sand and desert dunes. London: Methuen 1941.

[20] Nalpanis, P.: Saltating and suspended particles over flat and sloping surfaces. II. Experiments and numerical simulations. Proc. Intern. Workshop Phys. Blown Sand, Aarhus, May 28—31, 1985, (ed. O. E. Barndorff-Nielsen) Dept. Theoretical Statistics, University of Aarhus, Denmark, pp. 37—66 (1985).

[21] Rasmussen, K. R., Mikkelsn, H. E.: Wind tunnel observations of aeolian transport rates (this volume).

[22] Rasmussen, K. R., Sørensen, M., Willetts, B. B.: Measurements of saltation and wind strength on beaches. Proc. Intern. Workshop Phys. Blown Sand, Aarhus, May 28—31, 1985 (ed. O. E. Barndorff-Nielsen) Dept. Theoretical Statistics, University of Aarhus, Denmark, pp. 301—325 (1985).

[23] Anderson, R. S.: Sediment transport by wind: saltation, suspension, erosion and ripples. Ph. D. thesis, University of Washington (1985).

[24] Sørensen, M.: Estimation of some aeolian saltation transport parameters from transport rate profiles. Proc. Intern. Workshop Phys. Blown Sand, Aarhus, May 28—31, 1985 (ed. O. E. Barndorff-Nielsen) Dept. Theoretical Statistics, University of Aarhus, Denmark, pp. 141—190 (1985).

Author's address: M. Raupach, CSIRO Centre for Environmental Mechanics, Black Mountain, GPO Box 821, Canberra, ACT 2601, Australia

Acta Mechanica (1991) [Suppl] 1: 97—122
© by Springer-Verlag 1991

Grain transport rates in steady and unsteady turbulent airflows

G. R. Butterfield, London, United Kingdom

Summary. Wind tunnel and field experiments are reported in which continuous, synchronous measurements of grain transport rates and near-bed velocity profiles were made at one second intervals to assess mass-flux response to velocity variations. Resulting grain flux and velocity series demonstrate the variability concealed by conventional time-averaged data. In steady tunnel winds, time-dependent mass transport rates are found to correlate better with fluctuations in mean velocity near the top of the saltation layer than with estimates of instantaneous shear stress. Quasi-periodic oscillation (20—30 seconds) of near-bed mass-flux and flow velocity in the lower regions of the inner boundary layer is evident in such airflows as the saltation system moves towards equilibrium with a developing bed form and confined boundary layer. This phenomenon may not occur in nature at these time-scales, however.

In systematically unsteady airflows, the time constant between flux rate and velocity near the top of the saltation layer is shown to be of order one second, tentatively confirming Anderson and Haff's [2] calculations of saltation response time. Mass-flux also correlates well with large time-dependent variations in velocity in this region. Without grain replenishment, progressive surface re-sorting induces non-stationarity in grain flux under all observed flow regimes. Mass-flux and velocity histories measured on dunes show no correspondence. This difference is attributed to the stochastic nature of three-dimensional turbulence, the larger integral scales of atmospheric flows, measurement noise, and the effects of flow non-uniformity on satisfactory definition of shear velocity. Unsteady velocity profiles over a transverse dune are shown to be non-logarithmic above 20 cm but log-linear velocity segments of variable extent are found within the upper saltation layer.

1 Introduction

Progress in modelling grain transport rates in air is currently constrained by a lack of reliable simultaneous measurements of wind characteristics and sand transport rates [3], and by use of time-averaged data that conceal the short-term interactive processes by which steady state transport and grain flux responses are achieved. For developed, steady-state saltation in air, available empirically derived models show time-averaged rate of total grain transport, q, is non-linearly dependent upon some time-averaged characteristic velocity, usually the shear (friction) velocity, $U*$, measured above the saltation layer [13], [17]. Despite refinements to include the effects of grain size, grain shape, density ratio and moisture levels, mass-flux relations of this kind generally show only moderate mutual agreement [21], [28], [33], [34]. Significantly, such relations have usually been calibrated using steady tunnel winds, boundary layers of uniform properties, and sand supply adjusted to maintain a constant bed surface condition considered typical of stable sand flow from extensive source areas.

However, under natural conditions such as those prevailing over dune slopes, wind velocities are commonly unsteady and often rapidly fluctuating giving an inner boundary layer of highly variable properties [39], and progressive re-sorting of surface grains may occur where source areas are of relatively restricted extent or exposure. In consequence, variations in grain transport rate might be expected to occur over periods of seconds, and over longer time periods non-stationary average flux rates are also probable. The relationship between grain transport rate and wind velocity at time-scales of less than a few minutes has not, however, been established for steady airflows, and mass-flux response to fluctuating and unsteady winds has not been assessed directly. This paper reports a series of wind tunnel and field experiments in which continuous, synchronous measurements of grain transport rates and near-bed wind velocity profiles were made at one second intervals in order to assess the effects of time constraints on mass-flux relationships.

Investigation of short-period variations in mass-flux and velocity offers a number of benefits. First, it permits assessment of response times for regulation of transport rate in steady-state saltation under constant winds. Since the influential papers of Bagnold [4] and Owen [26] established equilibrium saltation at a steady wind velocity has usually been assumed to be a self-regulating condition maintained by interactions between wind velocity alterations, air-borne grain concentrations and dislodgement rates. It is generally assumed that transport rate adjusts very quickly to flow changes [e.g. 2, 31], and numerical simulations by Anderson and Haff [2] suggest that response time for a saltation system to reach steady-state is roughly one to two seconds, but there are currently no published experimental data to test these assertions. Secondly, time-series of mass-flux in saltation under steady winds permit analysis of non-stationarity in grain transport rates and of temporal aspects of sand trap behaviour. Such non-stationarity may arise during bed adjustment to newly imposed flow regimes, in developing saltation over leading edge deposits [27], or where changes in surface erodibility occur through time as a result of re-sorting or changes in ambient surface moisture regime. Finally, if resolution of grain flux is adequate, the more complex case of grain transport in naturally fluctuating airflows may be analysed.

2 Instrumentation and field sites

In order to measure near-bed velocity profiles and short-period grain transport rates in a range of steady and unsteady winds under field conditions, robust hot-wire anemometers and a sand trap fitted with a continuously recording, sensitive load cell were used. Field experiments in natural winds were conducted using this apparatus on interdune and intertidal sands at Ynyslas, Wales, and in the crestal regions of transverse coastal dunes at Praia Bordeira, SW Portugal. To achieve the required control of airflow variation, and to enable transport rates to be determined for a range of undisturbed natural sand surfaces, a portable wind tunnel of the blow-down type was used over *in situ* dune sands at the Ynyslas site. Subsequently controlled experiments were conducted in the laboratory using the tunnel erected over a 10 cm deep bed of dune sand from the same site.

2.1 Measurement of short-period mass-flux

After preliminary feasibility tests, a sensitive electronic load cell with a resolution of 1 mg was fitted beneath two types of surface-mounted sand trap. Despite operational constraints imposed by the use of traps and pressure sensitive devices, this method was considered the most expedient way in which short-period grain fluxes might be consistently measured by a device usable in both natural and tunnel winds. For load increments ≤ 0.5 g, the load cell shows minimal resonance and has a calibrated response time of < 0.5 seconds which constrains resolvable mass-fluxes to those of order 0.5 g cm^{-1} s^{-1} or less. It is powered by either AC mains or batteries, and may be interfaced to a portable data logger or printer. Sand captured by the trap is funnelled directly into a screened container which collects the accumulating load (Fig. 1a).

Due to their simplicity and low-cost, most studies of aeolian sediment transport employ traps to measure sand transport rates, [e.g. 6, 14, 21, 41], and are, in consequence, constrained by trap efficiency. This is also true in this investigation. (Detailed discussion of the efficiency of sand traps may be found in Rasmussen and Mikkelsen [30], and Phillips and Willetts [28]). However, for reasons of practicality, cost and comparability with existing data, a continuously recording sand trap was considered preferable to the more sophisticated piezoelectric impactor devices of the type developed by Gillette and Stockton [12]. Truly isokinetic grain capture can only be achieved using artificial ventilation of traps and this is rarely practicable for field installations and is incompatible with load cell operation.

In this investigation uni-directional, naturally ventilated, wedge-shaped traps were fitted over the load cell for determinations of short-period grain flux. In initial wind tunnel runs, and in the majority of natural wind experiments, a perspex 'total load' trap with a segmented aperture 1 cm wide and 30 cm high and body flaring to 10 cm over a length of 26 cm was used to measure incremental total sand transport. The design of this trap derives from those of Horikawa and Shen [16] and Greeley et al. [14] but modified by addition of streamlining to reduce flow impedance, a subsurface sloping tube to duct creep and reptation loads directly to the load cell, and later, by horns (4 cm long) protuding either side of the front aperture. In this design flow stagnation is reduced (but not eliminated) by a substantial throughflow of air and by the pressure differential created within the trap by its flared shape. In natural boundary layers a trap of this vertical dimension was needed, but for use with the wind tunnel it proved unnecessarily large and was used only for comparative purposes. In the field excavation behind a retaining plate permits installation of the load cell unit below the sand surface level (Fig. 1a). The trap is then fitted to the load cell unit so that the lip of the aperture rests close to the undisturbed sand surface. Directional adjustment within an arc $\pm 30°$ is then made to align the trap to the wind direction indicated by a small vane mounted on the trap, and a laboratory jack is used to adjust the elevation of the trap and cell to ensure perfect continuity of the aperture with the sand surface. Finally, sand is replaced around the entire unit. A taring facility allows the weight of any sand admitted during installation to be deducted and consecutive test runs to be initiated from a distant control station without further disturbance of the trap or sand surface.

In operation on dry sand, scour develops about the body of the 'total load' trap and progresses upwind towards the aperture. After 120 to 180 seconds, depending on velocity, bed continuity with the lip of the aperture ceases and capture efficiency declines sharply. Under steady winds this point is readily discernible in recorded time-series of grain flux

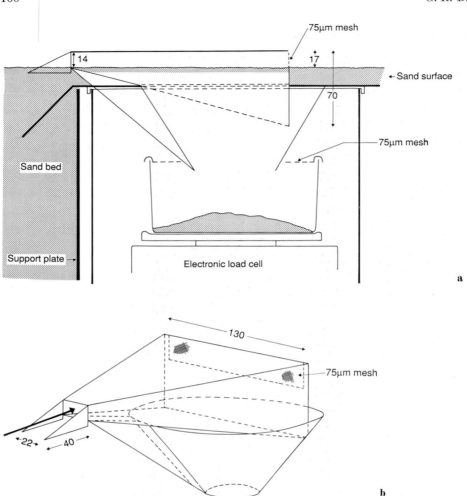

Fig. 1. Schematic view of apparatus for measurement of short-period grain flux. (All dimensions in millimeters.) **a** Reptation trap and load cell configuration used in wind tunnel runs. (Field installation is identical except for use of 'total load' sand trap.) **b** Dimensions of reptation trap

and this, together with visual observations, enables runs to be terminated before capture efficiency is seriously impaired. In practice all runs with this trap were limited to periods of 180 seconds or less before any scour was infilled manually and the trap elevation was re-adjusted.

However, as the 'total load' trap could not be calibrated and induced unacceptable scouring in prolonged use, in later tunnel experiments the efficiency of near-bed capture was greatly increased by use of a modified form of the nearly isokinetic 'reptation trap' designed and calibrated by Rasmussen and Mikkelsen [30]. This trap, which is shown in Fig. 1b, was built to the specifications of these authors but for two modifications: (1) the aperture dimensions (22 mm wide × 14 mm high) were reduced to restrict mass capture per unit time under high U_* values to amounts compatible with load cell resolution, and (2) a sloping duct section was added to convey slow moving grains directly to the load container. The calibrated efficiency of this trap is 0.75 to 0.80 for U_* between 0.30 and 0.85 m s^{-1} [30] and in the present investigation no scour effects were observed about the trap even in

Fig. 2. View of reptation trap and bed condition in the portable wind tunnel after 10 minute exposure to a free-stream velocity of 7.62 m s^{-1}. Bed of Ynyslas sand (mean grain size 0.177 mm) with mean ripple length 6.4 cm

tunnel runs lasting up to 10 minutes during which ripples were seen to pass the trap unhindered (Fig. 2). Although sampling only the intense near-bed mass fluxes, the reptation trap mounted over the load cell unit undoubtedly provided the most reliable short-period flux data gathered in this investigation.

Prior to use, both traps were calibrated for the effects of wind force by recording cell load response to a range of applied steady flows blowing over fixed, rippled surfaces in the wind tunnel. Calibrations for the 'reptation' and 'total load' traps were made against velocities measured at a heights of 2 cm and 10 cm, respectively.

2.2 Measurement of mean mass-flux profile

Average vertical mass-flux profiles were measured using ventilated, chambered traps exposed for an appropriate time during each run. In the open, a modular trap consisting of 15 metal tubes each 20 cm long was used to catch grains up to 42 cm from the surface. Each tube has lateral ventilation through a 90 μm mesh, slopes downwind at an angle of 30°, and collects through an aperture 22 mm wide and 25 mm high. The lowermost chamber is placed flush with the surface and extends below it. The catch is recorded by weighing each chamber before and after a run. This uni-directional trap was installed for each run and was never exposed for longer than three minutes during which time scour about the lowest chamber remained within acceptable limits. In the thin boundary layer of the wind tunnel, traps of the 'Aarhus type' [19], [30] made from twin-wall, polycarbonate sheet were used to collect grains in eight vertical chambers each 8.5 × 8.5 mm cross-section and 200 mm long. Wall thickness is only 0.5 mm and mesh ventilation is provided on both sides and rear of the trap. In this case traps were held flush with the sand surface for periods ranging from 15 to 30 seconds only, depending on flow velocity, as longer exposures resulted in markedly declining capture efficiencies in the lowest chambers. Operated in this manner, these traps also provided increased resolution of near-bed mass-flux profiles in natural saltation during field measurements.

2.3 The portable wind tunnel

The portable wind tunnel is shown schematically in Fig. 3. The tunnel is of modular construction, made from aluminium and steel sheeting, and designed for ease of transportation and operation by two persons. A variable pitch axial flow fan 45 cm in diameter driven by a 1.1 kW motor powered from a generator blows air through a flow straightening section including honeycomb and up to three removable gauzes. Flow then enters a contraction section before emerging into a 460 cm long tunnel of uniform cross-section 30 × 30 cm. Each tunnel section is made from three hinged aluminium panels each with inset perspex windows and linkage units. In the field the blower, flow-straightening and contraction units are bolted together against seals, and the tunnel sections are carefully aligned and placed over the natural sand surface. Sharp side wall skirts pressed into the sand ensure a good seal against the natural bed, and the sections are sealed with tape and secured with cross frames driven into the sand. The fan runs at constant speed and flow rate is controlled by air-bleed throttles on either side of the flow straightening unit. Increased velocities can be achieved by selective removal of the gauzes but with some loss of flow stability, therefore this facility was not used in the present investigation. With all gauzes in place laboratory calibration shows flow within the tunnel is relatively uniform spanwise, except for the regions within 3 cm of the side walls.

The short length of the tunnel, a limitation imposed by portability and motor power, restricts boundary layer thickness to about 10 cm or less depending upon free-stream velocity and, over most and surfaces, probably constrains development of equilibrium grain transport. In the experiments reported here, no sand feed was used as initial experiments

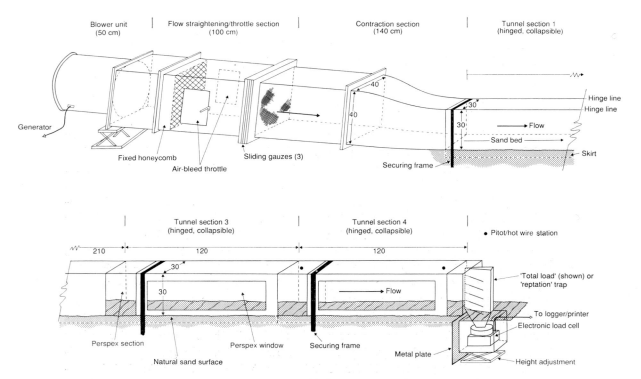

Fig. 3. Schematic view of portable wind tunnel equipped for short-period grain flux monitoring in the field. (All dimensions in centimetres.)

using a sand feed device showed measured short-period grain fluxes to be extremely sensitive to rates of upwind sand supply. Despite these limitations, the tunnel provides a controllable boundary layer flow in which relative transport rates can be determined for a range of steady and unsteady airflow conditions over undisturbed sand beds, and simulates reasonably conditions of developing saltation.

2.4 Airflow measurement

Natural wind velocity profiles between 15 and 320 cm were measured over dunes, interdune corridors and intertidal flats using a modified Thornthwaite Wind Profiling System Model 106. This consists of six sensitive, matched cup anemometers (cup assembly weight 7 g; distance constant 83.3 cm) that may be mounted at variable heights on a single mast (Fig. 4) or singly on individual masts located up to 60 m distant from the control unit. Time-aver aged wind velocities over selected integration times in excess of one minute are registered by LED's and in this study were recorded verbally by dictaphone. Each unit was calibrated by the manufacturer and subsequently in a large wind tunnel during these experiments. Temperatures were measured with three Maplin digital thermometers (resolution 0.1 °C) mounted at 10, 50 and 100 cm height.

In the open and in the tunnel wind speeds within 10 cm of the surface were measured using an array of four Dantec steel-clad hot-wire probes giving horizontal velocities integrated over averaging times ranging from 1 to 180 seconds. Synchronized output was recorded using portable printers and data loggers. In the open, velocities were measured at four heights between 2 and 10 cm (determined to ± 2 mm) with the lower probes (4 mm diam.) embedded vertically in the sand, and the upper ones (7 mm diam.) usually suspended

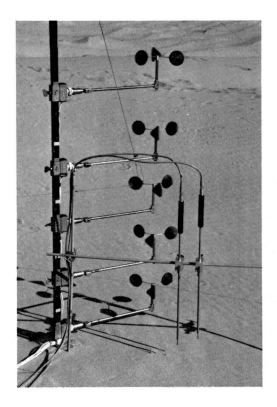

Fig. 4. Instrument array used for field determination of boundary layer velocity profile characteristics. Between $z = 2$ to 10 cm, array comprises four steel-clad hot-wire probes and two temperature sensors; between 10 and 150—320 cm six sensitive cup anemometers and three temperature sensors are deployed

vertically at variable heights from an 'H' shaped frame (Fig. 4). In the tunnel, mean boundary layer velocity profiles within 10 cm of the bed were determined by vertical traverse of a fine pitot tube and a pair of hot-wire probes, all positioned to ± 1 mm using a vernier racking system. Short-period velocity profile measurements were made within the same region using two pairs of hot wire probes held in a fixed array. As discussed below, however, the design and dimensions of the hot-wire probes render them unreliable close to the bed in thin boundary layers or intense shear flows. Hence, probe performance was carefully calibrated against velocities measured using fixed-height pitot combs in wind tunnel boundary layer flows, and only tunnel velocities measured by hot-wire probes positioned above 2 cm from the surface are used in analyses presented here.

The flexibility and portability of the anemometry systems used in this study permitted mean boundary layer velocity profiles to be determined in the field over periods of 60 seconds or longer at any elevation between 2 and 320 cm, whilst synchronized time-series of one second average velocities could be obtained from up to four hot-wire probes measuring below 10 cm from the surface. In the tunnel, flow characteristics within the 'inner boundary layer' above the saltation layer [see 27], and to a limited extent within the saltation layer itself, could be ascertained for correlation with measured short-period grain fluxes.

3 Experiments and procedures

Two series of experiments are reported here. First, wind tunnel experiments in which (1) steady, (2) systematically unsteady and (3) randomly fluctuating free-stream velocities were applied to beds of interdune sands from the Ynyslas site, both *in situ* and in the laboratory, and one second values of both grain flux (q') and boundary layer flow velocities (U') were measured. Secondly, field measurements in which wind profile characteristics and q' were measured under natural, unsteady winds over dune crests at Praia Bordeira and intertidal flats at Ynyslas. The grain size distributions of the interdune sands from Ynyslas (mean size 0.177 mm) and the crestal dune sands from Bordeira (mean size 0.270 mm) are shown in Fig. 5. Using the classification of Folk and Ward [10], both sands are very well sorted.

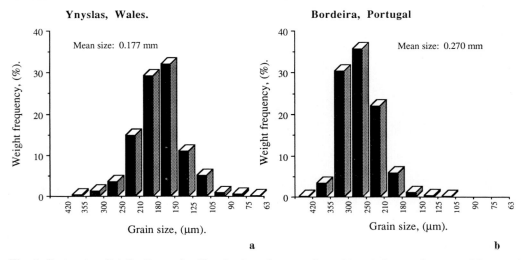

Fig. 5. Grain size distributions of **a** Ynyslas interdune sand used in wind tunnel runs and **b** crestal dune sands from the Bordeira field site

3.1 Wind tunnel experiments

Wind tunnel experiments were conducted initially in the field over undisturbed interdune sand and subsequently in the laboratory over a well-mixed bed of the same sand smoothed level with the sides of a 10 cm deep sand-retaining box to which the tunnel was fixed. In all cases the bed was first allowed to establish an equilibrium surface form by pre-exposure for several minutes to the lowest intended velocity. To prevent excessive upwind scour and prolong the life of the test bed, in some cases it was necessary to fix the surface of the leading 50 cm of the established bed by spraying with water. In the absence of a regulated sand feed, this also allowed sand transport to develop across a homogeneous bed surface. The trap ('total load' or 'reptation') and load cell were then installed at the tunnel exit (Fig. 3), tested and adjusted. In initial experiments a strain gauge turbulence probe [8] was fixed to the tunnel roof to measure fluctuations in tunnel reference velocity 15 cm upwind of the trap aperture, but for the majority of experiments arrays of two or more hot-wire probes were mounted on a vernier controlled racking system in this position allowing velocity profiles and U_* to be determined.

For steady winds, each experimental run was initiated by running the tunnel at a pre-determined velocity for 15 seconds and then simultaneously taring the load cell and triggering the synchronous recording each second of cumulative cell load and velocity values from all hot-wire probes. Careful prior adjustment of interface scan rates ensured synchroneity in recording of cumulative cell load and flow velocities. Cumulative cell load was later corrected for dynamic wind loadings using the calibration relations and hot-wire velocity measurements from the appropriate height. Incremental dry sand weights thus yielded grain transport rates per second without further adjustment. For unsteady flows, runs were always initiated at a low, steady velocity, and cumulative cell load arising during subsequent faster flows was later adjusted for the calibrated wind force effects of the corresponding measured velocity. Unsteady and randomly fluctuating flows were achieved by manual manipulation of the tunnel air-bleed throttle.

Typically all runs were limited to 120—180 seconds duration to ensure consistent trap efficiency, prolong bed life and restrict data acquisition to manageable levels. If required the captured load was recovered upon completion of the run, but more commonly trap elevation was adjusted, the bed in the vicinity of the trap was reconstructed, and the next run was initiated by repeating the procedures outlined above. During selected runs the 'Aarhus' trap was used to measure the transport rate profile.

Two kinds of velocity data were gathered in wind tunnel runs. First, for each bed type and throttle setting the mean velocity profile 15 cm upwind of the trap was measured by traversing both a hot-wire probe and a pitot tube vertically from near the bed to the free-stream flow. Although use of an electronic micromanometer hastened this process, the time needed to achieve manometer stability restricted measurements to less than six heights in a three minute run. Profile traverses using racked pairs of hot-wire probes and an integration time of 10 seconds, however, allowed up to 14 measurements to be made at heights between 7 and 120 mm. Such data allowed mean velocity profiles during active saltation to be established, and calibration uncertainties arising from use of relatively large hot-wire probes on a thin shear flow to be checked. Secondly, up to four hot-wire probes were located at fixed heights within the internal boundary layer defined by the logarithmic velocity gradient discernible above the saltation layer in measured mean velocity profiles. Simultaneous measurements of velocity were taken at these heights each second on order to establish velocity series at time-scales identical to those of grain flux series.

3.2 Field measurements

In the second series of experiments, short-period grain fluxes and flow velocities were measured under natural winds. The 'total load' trap was fitted to the the load cell unit in excavations dug into dune crest sands or intertidal flats. Installation is difficult when sand is mobile and was best achieved at both field sites in calm periods before anticipated winds. Hot-wire and cup anemometer arrays were positioned alongside or 1 m upwind of the 'total load' trap unit, and both metal-tube and 'Aarhus type' vertical distribution traps were installed nearby. After a period during which the trap was allowed to 'bed in' under active saltation, consecutive runs were initiated in the manner described for tunnel runs. As under field conditions most winds are gusty, the load cell was tared with the trap aperture sealed. This allowed subsequent adjustment of cumulative load data using appropriate calibration factors for each one second wind velocity value measured simultaneously by a hot-wire probe placed at the standard height (10 cm) used in earlier calibration of wind force effects on the load cell.

4 Results and discussion

4.1 Grain flux in steady tunnel airflows

In this investigation steady wind cases are confined to wind tunnel runs over interdune sand from the Ynyslas site. Discussion in this section is therefore restricted to results from wind tunnel observations in the field and in the laboratory.

Figure 6 gives a comparison of typical mean velocity profiles, corrected for variations in air temperature, relative humidity and barometric pressure, measured by hot-wire and pitot traverses in steady tunnel flows over both fixed, planar and mobile, rippled beds of Ynyslas sand. All profiles show boundary layer thickness reaches no more than 8 cm at the tunnel exit with a good log-linear fit to the data in the region $2 \leq z \leq 8$ cm, while those measured during saltation show a clearly defined near-bed segment of apparently lower velocity gradient. Restricted boundary layer development is to be expected given the limited length of the tunnel, and lower near-bed velocity gradients (steeper slope as conventionally plotted in Fig. 6) are commonly reported for velocity profiles measured during saltation [9], [11], [41]. However, a tendency for decreased velocity gradients in the region $z < 2$ cm is also apparent in some fixed-bed profiles obtained by hot-wire probes. This feature is not evident in profiles measured by pitot tube and may arise from the effects of probe intrusion into the intense near-bed shear flow, or from imprecise definition of the height at which probes sense velocity. Naturally the latter effect becomes more critical near the bed. As a precautionary measure, therefore, hot-wire velocity measurements taken below 2 cm from the bed have been excluded from the analyses. At all other elevations mean velocities measured by hot-wire and pitot tube show excellent agreement validating velocity profile determination by the hot-wire probes.

It is considered that the log-linear velocity segment above about 2 cm from the bed represents the developing 'internal boundary layer' above the saltation curtain, as defined by Owen and Gillette [27] for flows over leading edge deposits. This is confirmed by analysis of transport rate profiles from tunnel runs (Fig. 7) which typically exhibit two log-linear segments: a lower segment below 1.8 cm from the bed in which, on average, 79%

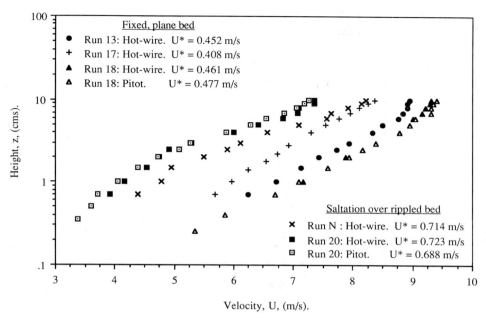

Fig. 6. Mean wind-velocity profiles measured in the wind tunnel by hot-wire and pitot traverses over fixed, plane beds in absence of saltation, and during active saltation over beds of developed ripples. (Sand 0.177 mm in all cases; open symbols denote velocity determination by pitot tube.)

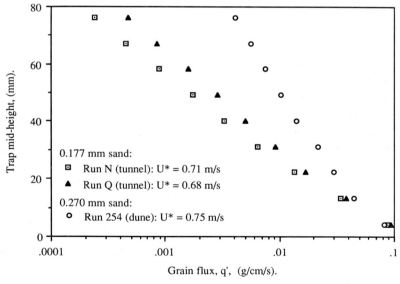

Fig. 7. Examples of near-bed transport rate profiles measured using 'Aarhus type' traps in the wind tunnel and on a dune crest. Each profile represents mean horizontal grain flux determined from three 30 second periods within a three minute run

of grain transport occurs and sand flux increases rapidly towards the bed, and an upper segment of lower flux gradient representing the remaining load. Grain fluxes measured by 'reptation trap' (vertical aperture dimension 1.4 cm) are, therefore, confined to the lower region of steep flux gradient. Interestingly, a similar partitioning of the transport rate profile at about the same height is discernible in profiles measured under natural winds (c.f. Fig. 14). The double log-linear profile accords well with the findings of Kawamura [22] and Rasmussen and Mikkelsen [30] and supports the evidence of mean velocity profile data (Fig. 6) indicating a region of grain-dominated flow below about 2 cm from the bed characterized by significant grain-borne shear stress and overlain by a region of shear flow apparently largely unaffected by grain excursions.

Consequently, following Gerety [11], mean shear velocities (U_*) during saltation were calculated by log-law regression fit (95% confidence limits) to velocities measured in the constant stress layer between 2 and 8 cm above the sand surface. One second shear velocities (U_*') were usually evaluated for each run from hot-wire measurements at four heights within this layer using the same method but 90% confidence limits. In some cases, however, only three probes lay within the constant stress layer identified in mean velocity profiles thereby reducing the reliability of derived U_*' values. For each two or three minute run arithmetic mean and root-mean square (σ) were computed for all U', q' and U_*' series. In addition, the local longitudinal turbulence intensity (T) at each measurement height was calculated as σ_u'/U, where σ_u' is the root-mean square (rms) of velocity fluctuations in the mean flow direction measured at the same height as the mean velocity, U [35]. Optimum serial correspondence between time-series of q' and U_*' was tested using cross-correlation. In all cases in which boundary layer velocity profiles in the tunnel were measured within 15 cm of the trap aperture, best match was achieved with shear velocity lagged one second with respect to grain flux. This is consistent with the estimated effects of grain travel time within the trap, probe displacement upwind of the trap, and of the calibrated response time of the load cell.

Figure 8a shows synchronized series of one second mass-flux (reptation trap data) and computed shear velocity for a typical run under steady tunnel flow. The measured grain fluxes are over a mature bed of well-developed ripples with an average wavelength of 6.25 cm. Regardless of the free-stream velocity used, such time-series typically show considerable variation of both q' and U_*' over the measurement period and highlight the considerable range in these variables concealed by time-average values. However, some noise in U_*' series undoubtedly derives from the limited number of velocity measurement heights. Typically smoothing q' and U_*' series reveals that, after an initial period of system stabilization, progressive development of a quasi-periodic oscillation of mass-flux and shear velocity occurs showing an apparent inverse relationship between these two variables (Fig. 8b). As in this example, periodicity is variable at about 20—30 seconds, but generally there is reasonably good correspondence between the times of reversal in both series.

Analysis of velocity series for measurements within the internal boundary layer shows that with decreasing height σ_u' values increase markedly and the orderly coupling of velocities to those in the upper boundary layer expected in time-averaged flow appears to break down. In Fig. 9a, for clarity, only the velocity series at $z = 5$ cm (U_5') and $z = 10$ cm (U_{10}', effectively the free-stream velocity) are plotted for the same run shown in Fig. 8. Over the two minute period, the free-stream velocity shows relatively minor fluctuation ($T = 0.013$) about a non-stationary mean value declining some 0.15 m s^{-1}, whilst that measured at 5 cm shows considerably increased turbulence intensity ($T = 0.033$) but a stationary mean velocity. Transport rate profiles show flux rates at $z = 5$ cm in this run average

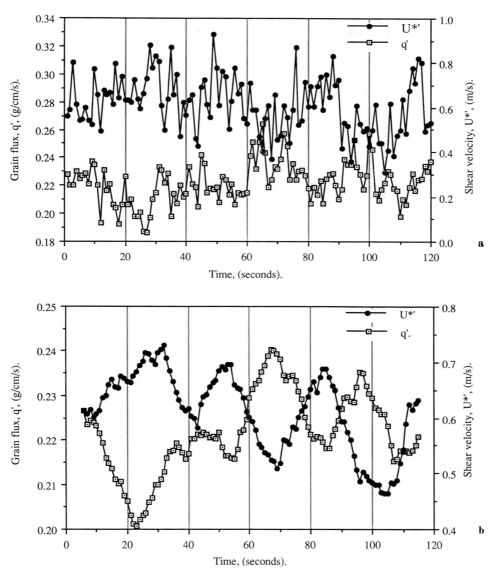

Fig. 8. Synchronized one second series of grain flux, q', (reptation trap data) and shear velocity, U_*', for steady tunnel flow (Run S25): **a** unsmoothed, and **b** smoothed ($x11$). Mean $q' = 0.222$ g cm^{-1} s^{-1}; coef. of variation = 6.59. Mean $U_* = 0.625$ m s^{-1}; $\sigma_{U_*}/U_* = 0.202$

2×10^{-3} g cm^{-1} s^{-1} with only 2.85% of total load transported above this height. Generally there is no correlation between variations of q' and those of free-stream velocity whether these data are smoothed or unsmoothed. However, in all runs there is weak correlation between q' and velocities at $2 \leqq z \leqq 5$ cm and, as expected, this improves considerably with data smoothing. Figure 9b shows that, for the smoothed data series of the example run, mass-flux variations under ostensibly constant airflow relate quite closely ($r^2 = 0.483$) to variations in velocity near the top of the saltation curtain. Given the noise intrinsic to the data, perfect correlation between such variables is most unlikely.

Results illustrated in Figs. 8 and 9 exemplify relations between grain flux and airflow in steady tunnel winds. First, it is clear that, once high frequency variations are removed

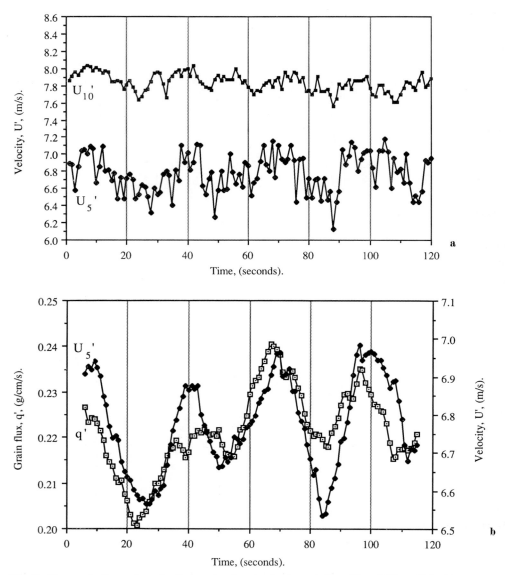

Fig. 9. Flow velocity and grain flux series for steady tunnel flow (Run S25). **a** Velocity at $z = 5$ cm (U_5': mean 6.78 m s^{-1}; $T = 0.033$) and $z = 10$ cm (U_{10}': mean 7.84 m s^{-1}; $T = 0.013$). Velocity U_{10}' is free-stream velocity. **b** Smoothed (x11) series of grain flux and U_5'

from the data, systematic covariation of near-bed mass-flux and flow velocity in the lower regions of the internal boundary layer is revealed. Secondly, under conditions of near constant free-stream velocity, smoothed one second grain transport rates measured within 14 mm of the surface decrease as shear velocity measured above the saltation layer increases. This apparently paradoxical condition seems to arise as, under nearly steady flow in the upper regions of the boundary layer, reductions in short-term mean velocity in the regions adjacent to the saltation layer accompany reductions in grain flux and induce increases in short-term mean shear velocity. A mutual interaction between near-bed grain flux and velocity in the lower boundary layer may be inferred, but loss of data resolution through smoothing prevents any indication that flux rate adjustments follow those of velocity or vice versa, and alternative explanations must also be considered. The include low-frequency fluctuations in tunnel speed, passage of bedforms and the effect of load cell response.

As periodicity and trends in smoothed series of free-stream velocity and grain flux show no correspondence, mass-flux variations on the time-scales seen here cannot be attributed directly to fluctuations in tunnel airflow. However, stationary airflow and sand transport conditions are difficult to achieve even in long wind tunnels [29], and Froude number dependent downstream variations in U_* are reported for developing saltation in small wind tunnels [27], [37]. White and Mounla [37] determined experimentally that a minimum entrance length corresponding to a tunnel length-to-height ratio of 5 is necessary to attain equilibrium saltation and constant values of shear velocity. This criterion is easily met by the dimensions of the portable tunnel, but applies only to the mean flow characteristics. It is possible, therefore, that some of the variations detected in smoothed U_*' series are due to natural instabilities in the saltation-flow system as it develops downstream towards equilibrium in the short working section of the tunnel. However, even if the observed temporal variations in U_*' and free-stream velocity are part of the natural evolution of the airflow and sand flux system towards some equilibrium state, the regulatory mechanisms are likely to be similar to those operative at steady-state even if the time-scales of their operation are different.

During most experiments, ripple migration occurred freely through the plane of the trap aperture. It might be speculated that variation in grain flux may have resulted from variations in local reptation and surface creep over ripples [1], [36] in the vicinity of the trap during their passage, and that near-bed flow acceleration in the wake region downwind of each ripple [23] could contribute to observed increases in velocity. However, measured ripple migration rates do not correspond to observed periodicities in either grain flux or near-bed velocities under steady flows, and the streamwise separation of velocity probes and reptation trap makes it unlikely that recorded variations would be in phase in any case. Finally, application of corrections for wind loading on the cell using 'best match' velocity measurements established by cross-correlation means that calibrated incremental cell loads should represent true grain loads and be largely unaffected by velocity variations through the trap aperture.

It is considered, therefore, that the observed modulation of near-bed grain flux and flow velocities in the lower regions of the inner boundary layer under constant tunnel winds probably represents a form of slow mutual self-regulation towards a state of dynamic equilibrium. However, in view of reported instability of U_* in developing saltation [29] [37], this process may only be a feature of incompletely developed saltation in a thin and developing boundary layer. Even so it has important implications for interpretation of sand transport experiments in wind tunnels.

4.2 Grain flux in unsteady tunnel flows

Grain flux response to simulated gusty winds can be assessed from tunnel runs in which velocity was varied systematically, randomly or a combination of both. When a velocity history contains constant and fluctuationg episodes, the response of the transport system to both prolonged and short-lived velocity excursions can be analysed. In the example reported here (Fig. 10), the tunnel throttle was manipulated to give periods at constant settings corresponding to pre-determined speeds separated by either instant changes in setting or

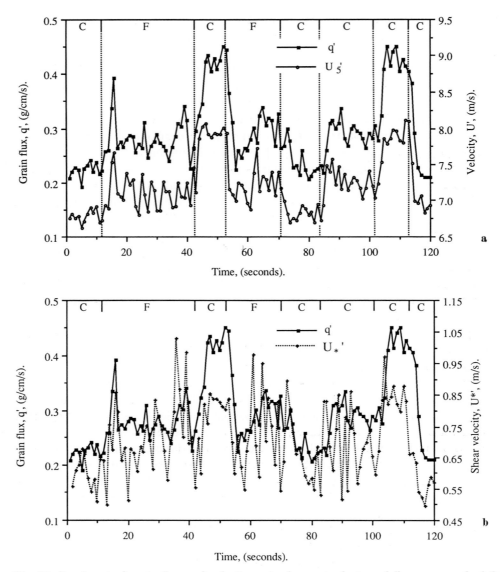

Fig. 10. Synchronized grain flux and velocity series for unsteady tunnel flow composed of fluctuating (F) and constant (C) periods. (Run S32: reptation trap data). **a** Grain flux and U_5' series. **b** Grain flux and U_*' series

random fluctuations within set limits. Although inertial effects induced by large pressure gradients almost certainly mean that large changes to throttle settings will not be instantaneously translated into velocity changes [cf. 32], the synchronous measurement of one second grain flux and velocity gradients in the vicinity of the trap means that this effect can be neglected here.

Figure 10 shows time-series of q', U_*' and U_5' in a systematically unsteady tunnel flow (Run S32). No adjustments have been made to match these series. Figure 10a shows that in sudden velocity increments to sustained higher speeds, grain flux responds within the one second resolution time of the instrument output. Increases in U_5' of order 1 m s^{-1} during active saltation take three seconds during which transport rate also rises. Once velocity begins to stabilize about the new value, grain flux also stabilizes, though there is a tendency for q' to peak after velocity in some cases. Flux reduction occurs immediately velocity begins to decline. In progressive flow accelerations and decelerations grain flux responds roughly linearly with U_*' and U_5'. As with steady wind cases, q' during periods of constant velocity at moderate to slow speeds is more closely related to velocities in the lower boundary layer than to U_*'; however, in periods of fast, steady wind, variance in U_*' is much reduced and reasonable agreement with grain flux variation is achieved.

In rapidly fluctuating airflows, recorded mass-flux seems unable to follow exactly the velocity variations in the lower boundary layer but, as expected, does mirror velocity trends closely. During a period of sustained velocity variation grain flux often builds progressively to a maximum. Significantly, U_*' ranges widely during periods of rapid velocity fluctuation at slow speeds, often giving 'spikes' well in excess of U_*' values associated with steady flow periods having a mean U_* of 0.15 m s^{-1} or more higher (Fig. 10 b). These features, which are not always matched by velocity peaks near the bed, sometimes induce matching spikes in the mass-flux history, especially when they occur in rapid succession.

Interpretation of time-series from unsteady flows must be qualified by the calibration uncertainties associated with the load cell response to rapid changes in applied wind force, and possibly sand load. Responses to uniformly increasing or decreasing loads during flow acceleration or deceleration are linear with neglible over-shoot, but, as the magnitude of high frequency pressure fluctuations increases, resonance and frequency modulation are likely but have not been assessed directly. Tested probe response, however, is sufficient to justify reliance on one second velocity data even in rapidly fluctuating airflows. Consequently, data series of the type shown in Fig. 10 are probably reasonable representations of events, except for grain flux data in periods of high turbulence intensity.

There is some evidence in the present data that at an established, roughly steady state, grain flux responds to the onset of a sustained increase or decrease in velocity within one second, though for the reasons given above a one second response time is indiscernible in rapidly fluctuating airflows. This may be taken as qualified confirmation of Anderson and Haff's [2] theoretical estimate that equilibrium between saltation and a newly imposed flow condition could be reached in one to two seconds. The observed tendency for reversals in sustained mass-flux rises to lag behind flow decelerations is probably attributable to the growing momentum of the intensifying saltation cloud prior to 'switch off'. Similar temporal cascades are reported for grain dislodgement and developing saltation [25], [40]. For the reasons discussed above, little of significance can be deduced from the poorly matched time-series of episodes of rapidly fluctuating airflow. It is clear, however, from the intensity of U_*' fluctuations at such times that flows in the upper regions of the boundary layer respond much more vigorously to throttle changes than flows nearer to the saltation layer. Consequently, short-lived sequences of shear velocity 'spikes' occur that are capable of inducing surges in grain transport during fluctuating flows without concomitant increases in velocity lower in the boundary layer.

4.3 Non-stationary transport rates in steady and unsteady tunnel flows

In this study, non-stationarity is a common feature of recorded grain flux histories where tunnel flows were imposed on pre-existing surfaces in the field, or where prolonged use was made of a test bed. Figure 11 shows an example of the former case. Here, grain flux is initially high due to disturbance of the surface during tunnel installation, but over the following two minutes decreases gradually by 0.25 g cm^{-1} s^{-1} as bed roughness adjusts to the newly imposed flow conditions. Mean velocity (U_5) shows a corresponding increase of order 0.08 m s^{-1} during this time as grain-borne shear stress decreases. Over the next 3.5 minutes transport rate and flow velocity remained stable but subsequently bed discontinuity with the trap aperture caused a rapid reduction in measured grain flux.

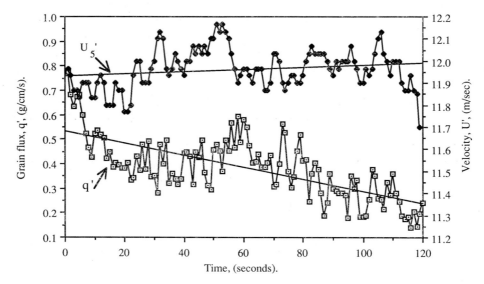

Fig. 11. Non-stationary transport rate (q') and velocity in lower internal boundary layer (U_5') during initial adjustment of transport system to flow regime imposed by portable wind tunnel on interdune surface, Ynsylas

No such problems were encountered in use of the 'reptation' trap yet measured transport rates in successive tunnel runs at a set velocity show progressive decline after periods of 8 to 12 minutes. Unfortunately the bed was not sampled during these runs but it seems probable that progressive re-sorting of the surface sand in the absence of a sand supply is the cause. Transport rate dependence on grain size has been reported previously [5], [20] and can lead to selective removal of grains with high transport potential, especially when, as here, there is no sand replenishment from upwind. Resultant modifications to roughness and grain availability then affect saltation reproduction rates and ultimately bring about non-stationarity in total transport rate. Such developments are also possible under natural conditions where upwind transportable sand is limited. This problem requires further investigation.

4.4 Grain flux in unsteady natural winds

Typical grain flux series measured within 30 cm of the surface under natural, unsteady winds are shown in Figs. 12 and 13b. The extreme variability of short-period grain flux under natural winds is well illustrated by the typical grain flux history shown in Fig. 12. In this example from Ynyslas, an onshore wind blew dry sand from the foreshore some 170 m across the damp surface of an estuarine intertidal flat before reaching the measurement point. Mean wind velocity and temperature gradients between 20 and 320 cm above the surface were measured using the cup anemometer array and shielded thermometers.

Stratification was found to be near-neutral so mean U_* was calculated from uncorrected wind velocities at each height yielding a value of 0.69 m s^{-1} for this run. The transport rate is characterised by low mean and high rms values, and rates range from near zero to in excess of 0.4 g cm^{-1} s^{-1} (see Fig. 12). Flux variability on this scale was never attained in wind tunnel experiments.

Photographic records show that the intensity of sampled saltation was not uniform spatially; semi-organised, sinuous, 'streaks' of more intense saltation with widths of order 0.2—1.2 m were seen to wander across the sand flat. As the position of such features is not fixed, the recorded sand flux history undoubtedly includes variations resulting from spatial variability of saltation, as well as from streamwise variations in windspeed. Sinuous streaks may be common features of all saltation clouds, but they are most commonly observed in saltation over flat terrain where structural features of the boundary layer are more likely to develop freely and find expression in the saltation cloud. These observations highlight the possibility that mass-flux histories resulting from Eulerian sampling under natural conditions may incorporate the effects of lateral variations in transport intensity. Three-dimensional aspects of transport rate variability are clearly of some significance to saltation rate modelling but as yet have not been researched.

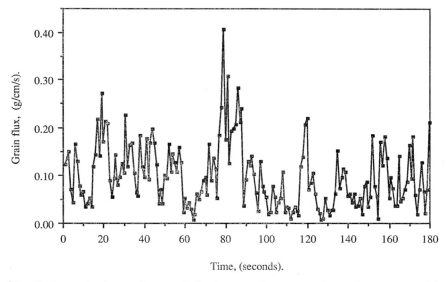

Time, (seconds).

Fig. 12. Example of grain flux instability in natural, gusty wind over damp, intertidal sands, Ynyslas. Grain flux measurement by 'total load' type trap; mean U_* during run: 0.68 m s^{-1}

Figures 13 and 14 show example results from measurements made near the crest of a transverse dune at Praia Bordeira (Run 248). The dune has a stoss face 42 m in length with an average slope of 8° but flattens to less than 1° over the crestal region. For the run shown, sand flux (Fig. 13b) and the transport rate profile (Fig. 14) were measured in the crest region 6 m from the dune brinkline, and wind velocities at 10 heights between 2 and 150 cm were measured 1 m upwind of the sand trap. Values of U_*' were calculated using log-law regression of velocities measured by hot-wire probes at four heights between 2 and 15 cm above the surface.

Even during periods of a few minutes, time-series of measured wind velocities and computed U_*' values from the dune crest are non-stationary, fluctuating, and show no discer-

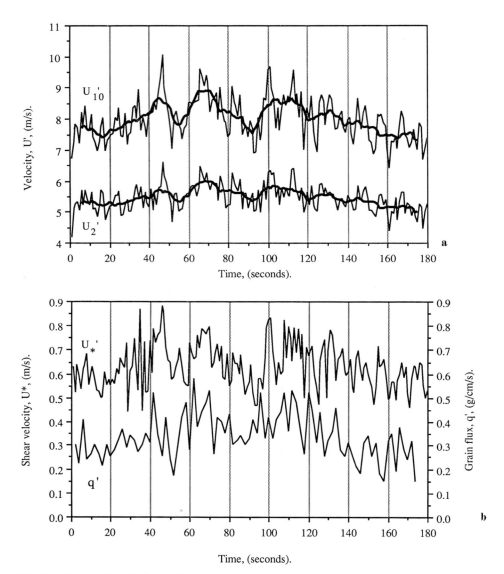

Fig. 13. Observed grain flux and velocity series (Run 248) on crest of transverse dune, Praia Bordeira. **a** Raw and smoothed hot-wire velocity series for heights of 2 and 10 cm within the saltation layer. **b** One second grain flux (q') and shear velocity (U_*') series, both unsmoothed and without cross-correlation adjustment.

(Mean $q = 0.336$ g cm^{-1} s^{-1}; coef. of variation $= 28.62$. Mean $U_* = 0.632$ m s^{-1}; $\sigma_{U_*}/U_* = 0.145$.)

nible correlation to measured q' values whether time-series are smoothed or unsmoothed (Fig. 13). Further, analysis of many such runs shows no unequivocal relationship between root-mean square of U' or U_*' and that of q'. This lack of correspondence is probably due principally to (1) the stochastic nature of natural, turbulent winds and their large integral scales, (2) data noise generated by the experimental arrangement and apparatus employed in the field, and (3) the effects of flow non-uniformity and saltation on definition of U_*.

Natural winds exhibit turbulence at a variety of time-scales [7], [15]. Episodic gusts over periods of minutes and quiescent periods of sub-threshold flows are easily recognisable in

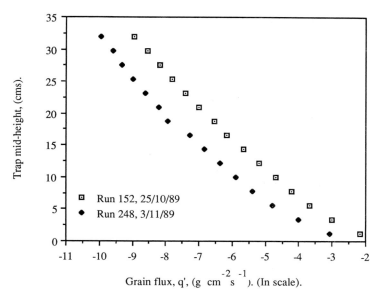

Fig. 14. Transport rate profiles at dune crest, Runs 248 and 152 (0.27 mm sand)

measured q' series, but the three-dimensional complexity of non-uniform flow at the dune crest effectively obscures simple relationships between sand transport rates and wind velocities over periods of seconds. In the example shown (Fig. 13), wind characteristics were measured 1 m upwind and 0.5 m to the side of the sand trap; in complex airflow over inhomogeneous dune topography such length scales are critical when short-period mass-flux relations are sought. Lags between change of velocity and corresponding changes in grain flux on dune slopes are expected [18] and will affect both time-averaged and short-period mass-flux relations, yet we have little knowledge of the appropriate length scales to apply in such situations. Partial explanation of the lack of correlation between short-period grain flux and air velocity lies in the large integral scales (lateral and longitudinal) in the atmosphere compared to the wind tunnel. Further, the uncertain response of the load cell to rapidly alternating pressures in highly turbulent airflows has already been commented upon, and load calibration to U' measured some distance upwind raises further uncertainties. Collectively these measurement problems increase as the averaging period for mass-flux relations decreases.

Finally, reliable determination of conventional mass-flux relations on dune slopes depends crucially upon appropriate definition of U_*. Estimation of U_* in uniform flows with saltation is itself problematic [11]. Over dunes flow acceleration strongly affects the turbulence structure of wind inducing non-uniformity and non-logarithmic velocity profiles [24]. Normalised mean velocity profiles, measured by cup anemometer arrays during active saltation and subsequently corrected for stability effects [31], are shown in Fig. 15 for four positions on a transverse dune at Praia Bordeira. Approximations to the unaccelerated velocity profiles based on velocity at 320 cm and estimated focal point parameters have been calculated for comparative purposes using the method of Mulligan [24]. Figure 15 confirms Mulligan's finding that the variable effects of flow acceleration over a dune slope makes derivation of U_* by conventional log-law fit to mean velocities measured above 20 cm during saltation events inappropriate.

To determine whether this problem might be overcome if velocities were measured below 20 cm, measurements were made using hot-wire probes. Resulting near-surface velocity profiles measured during saltation and normalised to mean velocity at 10 cm are shown in Fig. 16. From these it is apparent that, despite varying degrees of flow acceleration over the dune, an approximately logarithmic segment is identifiable in the region $6 \leqq z \leqq 20$ cm at locations on the lower half of the windward (stoss) dune slope and extends down to $z \approx 2$ cm in the crestal regions. For this reason, shear velocities at crestal dune locations were calculated from hot-wire velocity measurements between 2 and 15 cm from the surface, but at other locations this range would need to be modified.

The mean transport rate profile for Run 248 (Fig. 14) shows three log-linear segments separated by transitional regions at about 7 and 22 cm from the bed, and evidence from 'Aarhus' trap data (Fig. 7) indicates that a lower segment may also exist below about 2 cm. These segments most probably relate to grain trajectory height distributions [20], [30]. The segments between 0—7 and 7—22 cm in this case carried 85% and 13% of the total load, respectively, while 42% was transported within the 2—15 cm height range of hot-wire probe deployment. These observations confirm that measured velocity gradients are confined to the upper, less intense region of the saltation layer where velocity appears to be logarithmic allowing calculation of U_*. Whether U_* calculated in this region provides

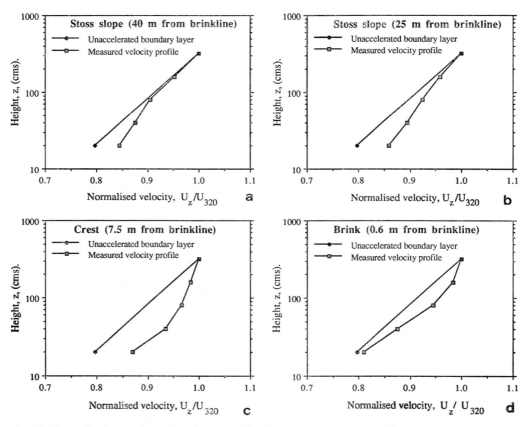

Fig. 15. Normalised mean boundary layer profiles for $z = 20$ to 320 cm at four points upwind of the brinkline on a transverse dune, Bordeira, 16 September, 1988. Velocities measured by cup anemometers (modified Thornthwaite Wind Profile System) and normalised to velocity at 320 cm. Unaccelerated boundary layer profiles approximate only (see text)

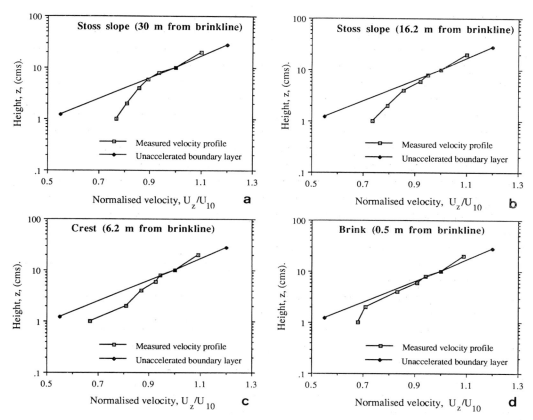

Fig. 16. Mean boundary layer velocity profiles for $z = 1$ to 20 cm at various locations on transverse dune, Bordeira, 3 November, 1989. Velocities measured by hot-wire probes and normalised to velocity at 10 cm. Unaccelerated boundary layer profiles estimated from focal point parameters (see text)

a more meaningful estimation of shear stresses regulating grain transport than that estimated from velocities measured in the higher regions of the profile is uncertain and requires further investigation. It is evident, however, that selection of an appropriate velocity profile segment in which to estimate U_* during saltation on dunes is not trivial and may contribute substantially to uncertainty in mass-flux relations on dune terrain.

5 Conclusions

Despite some operational constraints, use of robust hot-wire anemometers and a continuously recording sand trap has permitted useful preliminary analysis of relations between mass-flux and airflow at time-scales of seconds. With careful use and calibration, the sand trap apparatus has proven an expedient means of monitoring relative sand transport rates in steady airflows, and in those with sustained low frequency gusts, and the hot-wire probes have given reasonable resolution of velocity variations within and above the saltation layer. Measurement noise is evident in all data series derived in this study, and results probably raise more questions than answers, yet some tentative but important conclusions can be drawn.

First, grain flux under steady mean wind speeds is shown to be more variable than might have been expected from time-average values. Secondly, in the wind tunnel, time-dependent mass transport rates for steady mean winds appear to correlate better with fluctuations in mean wind velocity near the top of the saltation layer than with estimates of near instantaneous shear stress. Thirdly, mass-flux measurements in unsteady airflows correlate well with large time-dependent variations in velocity near the top of the saltation layer.

Fourthly, in developing saltation under steady tunnel winds, intense grain flux in a distinctive near-bed region of steep flux gradient appears to move progressively towards dynamic equilibrium with flow velocities in the lower part of an overlying inner boundary layer. As the quasi-periodic oscillation of mass-flux and velocity in the lower boundary layer evident in smoothed time-series is independent of velocity variations in the free-stream, it is concluded that this provides evidence for a mutual regulation of grain flux and airflow under these experimental conditions. Significantly, short-period shear velocity measured in the confined inner boundary layer of the wind tunnel is highly variable and, in the regulatory process, increases as grain flux decreases. Flux regulation of this kind may, however, scale with degree of boundary layer development and the consequent instability of U_* in the wind tunnel, and may not occur on these scales in developed atmospheric boundary layers. Extreme caution must, therefore, be exercised in extrapolating observed transport rate phenomena from wind tunnels into the real world.

Fifthly, results from unsteady tunnel flows show the time constant between mass-flux and velocity near the top of the saltation curtain appears to be of the order one second, similar to the calculation of Anderson and Haff [2]. Transport rates respond almost instantaneously to velocity accelerations, but lags of up to one second occur with flow deceleration. Despite uncertainties arising from the apparatus and techniques employed, such observations provide tentative confirmation that saltation systems respond to flow changes within one to two seconds. Shear velocity 'spikes' in rapidly fluctuating flows are not apparently followed exactly by grain flux changes but induce transport surges lasting several seconds.

Sixthly, establishing similar short-period mass-flux relations under field conditions is very exacting and, with the instrumentation employed in this study, has been relatively fruitless. The combined effects of time-lags, three-dimensional variability of sand transport, and turbulent velocity fluctuations at a variety of scales make this a problem best analysed by statistical methods. However, carefully designed field experiments using more equipment to deal with the larger integral scales of atmospheric flows may yet reveal correlations between short-period mass-flux and airflow conditions.

Finally, even time-average mass-flux relationships must be treated with caution given the problems of defining meaningful U_* for saltation in non-uniform airflows over irregular terrain. The notion of U_* is conventionally applied to *steady* wind where it defines a time-averaged rate of velocity increase with log-height. At time-scales of seconds, U_*' may have little physical meaning and, in any case, is difficult to define within acceptable confidence limits given a restricted number of velocity sensors. Transport rate profiles from tunnel and dune measurements show encouraging similarity, yet near-bed velocity profiles in saltation over dunes are much more complex than those in the wind tunnel. Much remains to be done to resolve the observed discrepancies between dune and tunnel observations. Further advances in the experimental investigation of short-period mass-flux relations seem, therefore, to require statistical analysis of Reynolds stresses, continuous monitoring of the grain flux profile by non-intrusive devices, and more extensive instrumentation to remove the restrictions of simple Eularian sampling.

Acknowledgements

The author gratefully acknowledges the financial support of the Royal Society (London) and the University of London Central Research Fund. The assistance and co-operation of many individuals and organisations is also much appreciated: Dr. M. Almeida (Lisbon) and P. Burden (Nature Conservancy Council, Ynyslas) for logistical support; Dr. T. Linsey and J. Purtill for assistance in the field. The helpful review comments of Professors J. D. Iversen and K. Hutter led to significant improvement of this paper. Above all sincere thanks to my wife P. Butterfield without whose practical assistance and continual patience and encouragement this research could not have been undertaken.

References

[1] Anderson, R. S.: A theoretical model for aeolian impact ripples. Sedimentology **34**, 943—956 (1987).

[2] Anderson, R. S., Haff, P. K.: Simulation of eolian saltation. Science **241**, 820—823 (1988).

[3] Anderson, R. S., Sørensen, M., Willetts, B. B.: A review of recent progress in our understanding of aeolian sediment transport (this volume).

[4] Bagnold, R. A.: The flow of cohesionless grains in fluids. Phil. Trans. Roy. Soc. **A, 249**, 235—297 (1956).

[5] Barndorff-Nielsen, O. E., Jensen, J. L., Nielsen, H. L., Rasmussen, K. R., Sørensen, M.: Wind tunnel tracer studies of grain progress. Proc. Int. Workshop on the Physics of Blown Sand, in Memoirs No. 8, Dept. Theor. Statist., Aarhus University, Denmark, **2**, 243—251 (1985).

[6] Belly, P. Y.: Sand movement by wind. Technical Memorandum No. 1, U.S. Army Coastal Eng. Res. Center, Washington D. C., 80 pp. (1964).

[7] Busch, N. E., Panofsky, H. A.: Recent spectra of atmospheric turbulence. Quart. J. Roy. Meteorol. Soc. **94**, 132—148 (1968).

[8] Butterfield, G. R.: The instrumentation and measurement of wind erosion. Proc. Sixth New Zealand Geog. Conf. **1**, 125—130 (1971).

[9] Chiu, T. Y.: Sand transport by wind. University of Florida (Gainsville), Dept. Coastal and Oceanographic Engineering, Technical Report TR-040, (1972).

[10] Folk, R. L., Ward, W. C.: Brazos River bar: a study of the significance of grain-size parameters. J. Sedim. Petrol. **27**, 3—26 (1957).

[11] Gerety, K. M.: Problems with determination of U_* from wind-velocity profiles measured in experiments with saltation. In: Barndorff-Nielsen, O. E. et al. (eds.): Proc. Int. Workshop on the Physics of Blown Sand, in Memoirs No. 8, Dept. Theor. Statist., Aarhus University, Denmark, **2**, 271—300 (1985).

[12] Gillette, D. A., Stockton, P. H.: Mass momentum and kinetic energy fluxes of saltating particles. In: Aeolian geomorphology, (W. G. Nickling ed.), pp. 35—56, Boston: Allen & Unwin 1986.

[13] Greeley, R., Iversen, J. D.: Wind as a geological process, p. 333. Cambridge: Cambridge University Press 1985.

[14] Greeley, R., Leach, R. N., Williams, S. H., White, B. R., Pollack, J. B., Krinsley, D. H., Marshall, J. R.: Rate of wind abrasion on Mars. J. Geophys. Res. **87**, 1009—1024 (1982).

[15] Høgstup, J.: Velocity spectra in the unstable planetary boundary layer. J. Atmospheric Sci. **39**, 2239—2248 (1982).

[16] Horikawa, K., Shen, H. W.: Sand movement by wind action (on the characteristics of sand traps). Tech. Mem. No. 119, 119pp, US Army Beach Erosion Board, Wash. D. C., 1960.

[17] Horikawa, K., Hotta, S., Kraus, N.: Literature review of sand transport by wind on a dry sand surface. Coastal Engineering **9**, 503—526 (1986).

[18] Howard, A. D., Walmsley, J. L.: Simulation model of isolated dune sculpture by wind. In: Barndorff-Nielsen, O. E. et. al. (eds.): Proc. Int. Workshop on the Physics of Blown Sand, in Memoirs No. 8, Dept. Theor. Statist, Aarhus University, Denmark, **2**, 377—390 (1985).

[19] Jensen, J. L., Rasmussen, K. R., Sørensen, M., Willetts, B. B.: The Hanstholm experiment 1982. Sand grain saltation on a beach. Dept. Theor. Statist., Aarhus University, Denmark, Research Report No. 125, 1984.

[20] Jensen, J. L., Sørensen, M.: Estimation of some aeolian saltation transport parameters: a reanalysis of Williams' data. Sedimentology **33**, 547—558 (1986).

[21] Jones, J. R., Willetts, B. B.: Errors in measuring uniform aeolian sand flow by means of an adjustable trap. Sedimentology **26**, 463—468 (1979).

[22] Kawamura, R.: Study on sand movement by wind. Reports of Physical Sciences research Institute of Tokyo University, 5, 95—112 (1951). [Translated from Japanese by National Aeronautic and Space Administration (NASA), Washington D. C. (1972)].

[23] McLean, S. R., Smith, J. D.: A model for flow over two-dimensional bed forms. J. Hydraul. Eng. **112**, 300—317 (1986).

[24] Mulligan, K. R.: Velocity profiles measured on the windward slope of a transverse dune. Earth Surface Processes and Landforms **13**, 573—582 (1988).

[25] Nickling, W. G.: The initiation of particle movement by wind. Sedimentology **35**, 499—511 (1988).

[26] Owen, P. R.: Saltation of uniform grains in air. J. Fluid Mech. **20**, 225—242 (1964).

[27] Owen, P. R., Gillette, D. A.: Wind tunnel constraint on saltation. In: Barndoff-Nielsen, O. E. et. al. (eds.): Proc. Int. Workshop on the Physics of Blown Sand, in Memoirs No. 8, Dept. Theor. Statist., Aarhus University, Denmark, **2**, 253—269 (1985).

[28] Phillips, C. J., Willetts, B. B.: A review of selected literature on sand stabilization. Coastal Engineering **2**, 133—147 (1978).

[29] Rasmussen, K. R., Mikkelsen, H. E.: Development of a boundary layer wind tunnel for aeolian studies. Geoskrift No. 27, Geologisk Institut, Aarhus University, Denmark, (1988).

[30] Rasmussen, K. R., Mikkelsen, H. E.: On the efficiency of sand traps and the transport rate profile. Sedimentology **38**, in press (1991).

[31] Rasmussen, K. R., Sørensen, M., Willetts, B. B.: Measurement of saltation and wind strength on beaches. In: Barndorff-Nielsen, O. E. et. al. (eds.): Proc. Int. Workshop on the Physics of Blown Sand, in Memoirs No. 8, Dept. Theor. Statist., Aarhus University, Denmark, **2**, 301—325 (1985).

[32] Raes, G.: Een windtunnelstudie over het effect van de tÿd op de granulometrie en de deflatiegevaeligheid by erosie van een 'loessig' materiaal. M. Sc. Thesis, K. U. Leuven, 116pp. 1988.

[33] Sarre, R. D.: Aeolian sand transport. Progr. Phys. Geog. **11**, 157—182 (1987).

[34] Sarre, R. D.: Evaluation of aeolian sand transport equations using intertidal zone measurements, Saunton Sands, England. Sedimentology **35**, 671—679 (1988).

[35] Task Committee on Preparation of Sedimentation Manual: Sediment transportation mechanics: wind erosion and transportation. J. Hydraul. Div. Am. Soc. Civ. Engrs. **91** (HY2), 267—287 (1965).

[36] Ungar, J. E., Haff, P. K.: Steady state saltation in air. Sedimentology **34**, 289—299 (1987).

[37] White, B. R., Mounla, H.: An experimental study of Froude number effect on wind-tunnel saltation (this volume).

[38] Williams, G.: Some aspects of the eolian saltation load. Sedimentology **3**, 257—287 (1964).

[39] Williams, J. J., Butterfield, G. R., Clark, D.: Aerodynamic entrainment threshold: effects of boundary layer flow conditions. Sedimentology **38**, in press (1991).

[40] Williams, J. J., Butterfield, G. R., Clark, D.: Rates of aerodynamic entrainment in a developing boundary layer. Sedimentology **37**, 1039—1048, (1990).

[41] Zingg, A. W.: Wind tunnel studies of the movement of sedimentary material. Proc. Fifth Hydraulics Conf., Iowa University Studies in Engineering Bulletin 34, 111—135 (1953).

Author's address: G. R. Butterfield, Department of Geography, Queen Mary and Westfield College, University of London, Mile End Road, London E1 4NS, United Kingdom

Acta Mechanica (1991) [Suppl] 1: 123—134
© by Springer-Verlag 1991

Initiation of motion of quartz sand grains

B. B. Willetts, I. K. McEwan, and M. A. Rice, Aberdeen, United Kingdom

Summary. Direct observation at a limited range of wind speeds of grain behaviour near the upwind edge of a sand deposit (by means of high-speed film) shows that disturbed grains usually, but not always, roll before taking off into reptation or saltation. Time-averaged description of the population of initial motions is deficient because of the importance of pronounced flurries of grain activity which occur at intervals. The flurries are clearly associated with flow features in the clean air wind because they occur too close to the leading edge to be impact generated.

An adapted saltation model is used to explore the development of activity downwind which results from a sequence of first dislodgements (into rolling or take-off) recorded on film. The portrayal of the development of a saltation layer is plausible and can be checked by means of flux profile measurements. However, it emphasises the need for more information about turbulence in the grain laden layer.

1 Introduction

It has long been accepted [1] that the grain dislodgements necessary to sustain aeolian transport of quartz sands are effected predominantly by inter-saltation collisions once the transport process is established. A good deal of knowledge about the collision process had been obtained in the last decade [2], [3], [4] and has been incorporated in numerical models of the saltation cloud which reproduce faithfully most of the observable features of the cloud [5], [6]. While such models contain well-founded accounts of the main features of high speed collision, they have hitherto extrapolated the collision data on the assumption that all new grain motions are the result of grain interaction. The success of the models is evidence that, at any rate in the cases so far modelled, collision is indeed the predominant disturbing agent.

Yet it is clear that the transport process must start with dislodgements caused by direct wind forces and furthermore there is no conclusive evidence that the role retained by such dislodgement becomes negligible when collision has become an important dislodging agent. There is some evidence that direct wind dislodgement retains considerable importance in transport of grains of low sphericity [7] for a large range of wind and grain circumstances. Thus for many circumstances, any model of grain behaviour clearly must include a procedure to portray the dislodgement of grains by direct wind forces. It can be argued that conditions deep in the grain laden layer are still so poorly understood that the existing models, apparently so successful for vigorous transport of compact grains, should yet include a direct dislodgement procedure. This would enable the assumptions made in them about reproduction rate of saltation (at present usually approximately 1:1) and about the death rate of reptations to be reviewed. Ideally, one should understand the variation of susceptibility to direct dislodgement as a function of grain size and shape and the near-bed velocity profile. However, this objective is a distant one and will certainly not be attained in this paper. When eventually such understanding is available, saltation cloud models will be adaptable to all grain types, and knowledge of the transport process will be greatly enhanced.

Unfortunately, the difficulties of studying direct dislodgement in the conditions of established aeolian transport are virtually insurmountable. Individual events are obscured by the cloud of flying grains, and it is not justifiable to create artificially sparse saltation clouds, as is done to study collision, because such sparse clouds do not affect the wind as does the cloud of normal grain concentration. Thus the wind retains a vigour (and an influence on the grains) which is not present in natural circumstances.

This study, acknowledging the difficulty of observing in the established grain cloud, was conducted near the leading edge of a sand deposit laid with its surface approximately co-planar with a flat floor of fixed roughness (equivalent to that of the mobile sand). Thus the wind contains no saltating grains when it reaches the deposit and its characteristics there are the well-known ones of fully developed flow over a flat rough surface. Upon encountering the moveable sand, it immediately begins to undergo changes, but it is very difficult to perceive any change of time-averaged velocity in the first 30 cm downwind of the deposit's leading edge.

The objective of the study was to observe the manner and frequency of direct wind dislodgement in the virtually sand-free wind, and to detect any changes evident in the few centimetres downwind of the leading edge. The direct applicability of the findings to the circumstances of active grain transport is limited. However, useful indirect application will be sought in two stages. The first is to study direct dislodgement as a function of wind structure. Then, after evaluating wind structure in the developed saltation cloud, a second difficult investigation, it will be possible to infer the direct dislodgement rate in those conditions. The inference will require cautious review because wind dislodgement and dislodgement by collision are not independent processes. Where there is a high intensity of collisions, the bed is in a generally disturbed state which makes direct dislodgement much easier than it is in the absence of saltation

2 Experimental procedure

Experiments were conducted in the Aberdeen wind tunnel which has a cross-section 500 mm $\times$ 500 mm and is of the blower type. Air is driven down it by a centifugal fan at the inlet end after passing through an expansion box and fine porous screens immediately downwind of the fan [8].

For the purpose of this experiment a rectangular recess was cut in the floor of the tunnel, as shown in Fig. 1. For several metres upwind of the recess the floor of the tunnel was covered with fixed roughness elements comprising grains from the population to be used in the experiment. The recess was then filled with loose grains of the same population, the deposit being levelled off almost flush with the general summit level of the fixed roughnes grains.

A high speed camera was mounted with optical axis very slightly above the plane of the stationary grain summits in the position indicated in Fig. 1. The length of the field of view 'a' was approximately 12 cm in most experiments, approximately 6 cm in others.

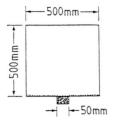

Section on A–A

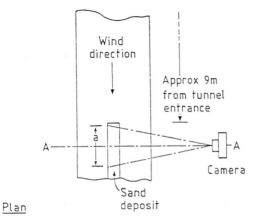

Plan

Fig. 1. Arrangement of the sand deposit in the wind tunnel and the camera position

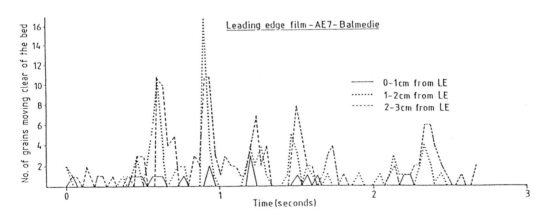

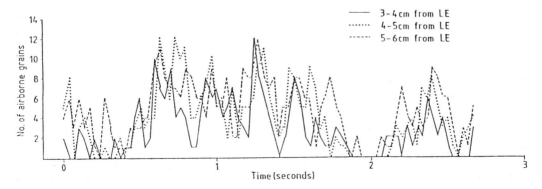

Fig. 2. Count of grains active as a function of distance from the leading edge and time. $u_* = 52$ cm/s

Air velocity was measured near the centre of the tunnel 150 cm upwind of the leading edge of the deposit at three heights above the base of the wind tunnel. At this point the wind is sand-free. The measurements were made using three pitot-static tubes connected to a diaphragm manometer in turn. Each measured velocity was thus time-averaged. From the local readings of velocity, values of u_*, the shear velocity, for the approach flow could be calculated reliably because the complications introduced by the saltation layer were absent at the site of the measurement. Experiments were done at four values of u_*: 43, 47, 50 and 52 cm/s. The impact theshold for this sand is approximately $u_{*t} = 24$ cm/s.

At each of these wind speeds, moving grains were filmed for a little over 3 seconds between 2.5 and 15 cm from the leading edge of the deposit, from the camera position shown in Fig. 1. The filming speed was 1 500 frames per second. The paths and velocity changes of moving grains in each film were then obtained by means of a digitising tablet. Each grain was followed through the field of view from first movement until it was lost to view beyond the downwind edge of the field of view. (The tracking was done in reverse for convenience). The start position relative to the leading edge was noted, together with the positions of first take-off into movement clear of the bed and of subsequent collisions with the bed. Velocities before and after each of these events were calculated.

One film (at $u_* = 52$ cm/s) with field of view extending for 6 cm from the leading edge was examined to determine the number of grains airborne in each of the first six 1 cm strips measured downwind (i.e. the strips 0—1 cm, 1—2 cm··· 5—6 cm downwind). This number was recorded as a time series, as illustrated in Fig. 2, for each of the six strips.

3 Results and discussion

Results were presented for the highest, 52 cm/s, and lowest, 43 cm/s, values of u_*. All show the undesirable raggedness which results from the small size of the sample (40 events in the first case and 43 in the second). Bearing this scatter in mind, there is no inconsistency between the results for these two wind speeds and those for the other two, which are not shown. These two will therefore serve to illustrate trends observable in all four cases.

Figure 3a shows that the number of grains rolling declines with distance from the leading edge beyond a value of about 7 cm in each case. On the premise that grains will generally hop rather than roll in established transport conditions, it is not unexpected to find the number of rolling grains declining as the transport develops.

However, Fig. 3b shows that the number of grains taking off from the bed, having increased between 4 cm and 8 cm at the higher wind speed and between 4 cm and 11 cm at the lower wind speed, also declines thereafter to low values at roughly 13 cm downwind. This is rather unexpected and suggests that it is possible that we may be seeing a transitory episode which, during the brief filming period, was showing particularly vigorous grain activity at this location. However, evidence that this is so is tenuous, and the possibility must be allowed that activity is more intense on average near the leading edge than it is 12 cm or so downwind.

In Fig. 4 the mean vertical component of take-off velocity is shown for each 1 cm strip for the two wind speeds. This variable increases reasonably progressively through the observed length of bed and slightly faster for the higher than for the lower wind speed. Thus, while the number of moving grains falls away in the interval 11—13 cm, the vigour of movement of grains there is certainly undiminished.

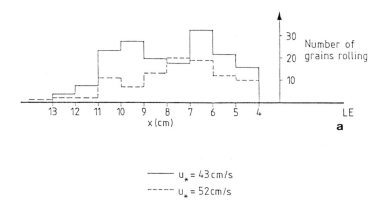

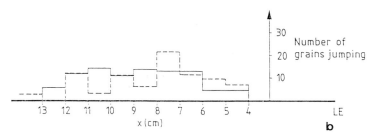

Fig. 3. The mode of transport of grains in the vicinity of the leading edge. **a** Number of rolling grains at two wind speeds. **b** Number of airborne grains at two wind speeds

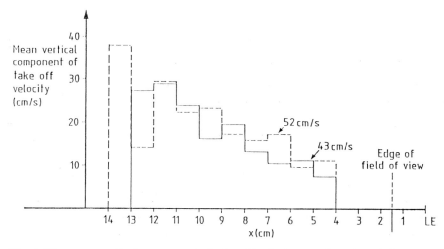

Fig. 4. The mean vertical component of take-off velocity as a function of distance from the leading edge, et two wind speeds

The way in which the mean value of the vertical component of take-off velocity increases is shown in a little more detail in Fig. 5. The proportion of observed grains falling into each of three bands of this variable is plotted for each cm from the leading edge. The bands are vertical component greater than (a) 15 cm/s, (b) 30 cm/s. and 50 cm/s. These correspond to trajectory heights exceeding approximately 1 mm, 5 mm and 13 mm respectively, indicating the range from feeble reptation to modest but recognisable saltation.

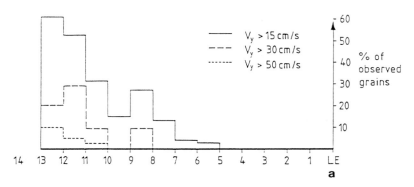

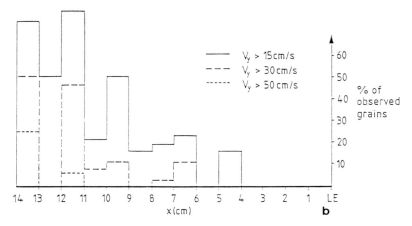

Fig. 5. The proportion of observed grains falling into each of three bands with respect to vertical component of take-off velocity. **a** at $u_* = 43$ cm/s **b** at $u_* = 52$ cm/s

Dependence on wind speed is obscured by the size and duration of the sample, but it is clear that trajectory height (as it is linked to vertical take-off speed) increases progressively from the leading edge. True saltation occurs at these wind speeds from about 10 cm downwind, although most movements there, certainly at the lower speed, fall well short of saltation.

An important additional consideration is evident on inspection of Fig. 2, which shows the variation of activity with time within 6 cm of the leading edge. While the trace for the strip 0—1 cm from the leading edge shows a low degree of activity with 0—3 grains in the air at any time, immediately downwind pronounced flurries of cativity are evident. Beween 1 and 2 cm, two intense but brief disturbances are seen in the latter half of the first second of the record. The number of airborne grains in each is an order of magnitude higher than it is in the remainder of the 2.5 seconds of the record. As one moves to stations farther from the leading edge the spikes in the record appear to be dispersed and perhaps phaseshifted.

It is clear that the disturbances at 1—2 cm are not related to upwind grain movements they are locally generated by time variations in the flow field. They themselves might be expected to prompt grain activity downwind as the disturbed grains are driven forward by the mean wind and embark on rolling/reptating/saltating careers. However, it is not evident whether the translating and dispersing activity observed downwind results from the consequences of the flurries seen between 1 and 2 cm, or from the action of a flow feature which itself translates and creates noticeable grain activity at a sequence of sites as it progresses. Both these effects may contribute significantly or one may be dominant.

4 A model of grain behaviour

The features of grain behaviour near the leading edge which have been reported above, reinforced by observation of the less easily quantifiable phenomena on film, suggest the following account of initiation of motion at this range of wind speeds. Most grains roll some distance before leaving the bed. The vigour of the first jump is related to the speed attained in rolling, and thence to the distance rolled. As more grains start to roll at x from the leading edge, they join a population already rolling after starts between the leading edge and x. Thus the rolling population grows with x until the number of conversions from rolling to jumping becomes equal to the number of rolling starts. This might be regarded as a description of time-averaged behaviour. However, superimposed on the general level of activity at each location, there are localised flurries of activity near the leading edge which are related to rather more persistent periods of activity at downstream locations. It seems probable that these flurries of activity are very important in the development of the stable transport rate downwind.

An exploration of this development was made using a model of grain activity which is reported elsewhere in this volume [9], with some adaptations, as will be described, to conform to observations near the leading edge. The basic model uses collision statistics to predict the outcome of each encounter with the bed, deterministic modelling of the grain path (ignoring turbulent features of the flow), and a grain-borne momentum extraction from the wind to produce a modified velocity profile for the grain-laden wind. However, for this application the wind profile modification routine is not used; grains are driven by a wind which remains unchanged by their presence. This is justified near the leading edge,

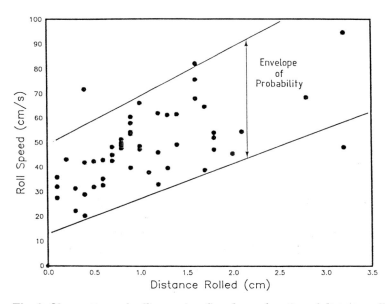

Fig. 6. Observations of rolling grains. Speed as a function of distance rolled from dislodgement

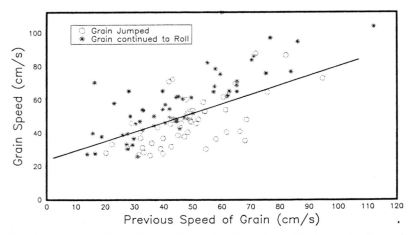

Fig. 7. The probability of take-off as a function of speed before and after bed contact. The indicated line demarks subsequent rolling from jumping in the calculation

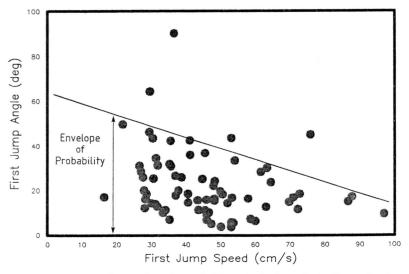

Fig. 8. Take-off angle *vs* take-off speed. The calculation selects the angle at random within the envelope of probability indicated, for a given speed

where the grain population is sparse. It becomes progressively less so as the population builds downwind towards a realistic uniform transport load. The simplified model consisting of trajectory and collision calculations was used as follows;

Observed initial movements from a leading edge film were entered at the appropriate place and time. Each may be classed as a roll or a direct take-off, depending on the magnitude of the initial vertical velocity component. Those which take-off follow one or more reptation/saltation trajectories, and are dealt with in the calculation as grains in a conventional saltation cloud model. Those which roll behave differently and require a new set of rules. The rules we have followed to predict their behaviour are these

1. For each rolling grain, the distance rolled, X_1, is recorded. Every 0.0001 second X_1 is updated and a random number is generated in the range 0 to 1. If the number is in the

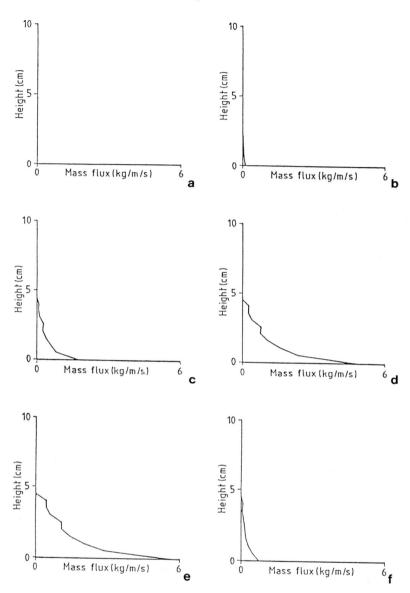

Fig. 9. The profile of flux with height calculated between 3.0 and 3.5 seconds from start-up at six positions: **a** 1 m **b** 3 m **c** 5 m **d** 6 m **e** 8 m **f** 10 m

range 0—0.99 the grain rolls on at unchanged velocity, but if it is greater than 0.99 the velocity is increased to a value which depends on X_1. This value is selected at random from the range indicated in Fig. 6 as a function of X_1. The new velocity may be a take-off speed or it may be in continued rolling. Figure 7, compiled from film data, is used to determine in which mode the grain continues. If it continues to roll, then this procedure is repeated for the next time-step.

2. If the grain takes off, then a take off angle is needed. This is taken by selecting randomly from the range shown in Fig. 8 for the speed obtained from the earlier procedure. Thus equiped with a take-off velocity vector, one can calculate the trajectory and its sequels. To save computation time, the take-off velocities were divided into a discrete number

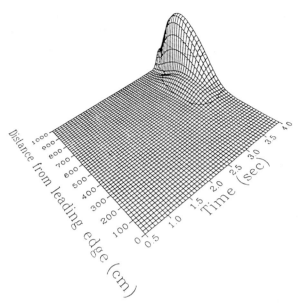

Fig. 10. Growth of mass flux of grains (kg/m/s) as a function of time and of distance from the leading edge

of classes, the representative trajectory for each of which was calculated only once. The sub-division favoured the smaller end of the trajectory range, so as to give better resolution for initial and immediately following movements.

3. The calculated trajectories and rolling sequence are computed synchronously, so that the position and speed of each grain is known at any chosen time. Many of the grains which are tracked, of course, are not those dislodged in the first 6 cm and 2.5 seconds, but are collision generated descendants of these. Nevertheless, all of the computed movements are consequences of wind-effected dislodgements in the first 6 cm.

The results of a computation following these rules are summarised in outline in Fig. 9 and 10. The input data are the first movements for the experiment represented by Fig. 2.

Figure 9 shows the development, with position, of the height at which grain activity occurs. This prediction is an interesting one because flux profiles are relatively easy to check experimentally. Therefore measured flux profiles could be freely used to validate improvements on our rather primitive first attempt to model the development of the saltation cloud.

Figure 10 summarises the transport rate changes in space (registered solely by distance from the leading edge) and time. It shows transport rate growing rapidly from 3 m downwind of the leading edge to a value at 7 m which is of the correct order for the wind condition in the experiment. This escallation of the transport rate appears to occur rather farther downwind than general wind-tunnel practice would suggest to be the common finding. However, the general features of the development of activity are reproduced quite accurately.

Without the "flurries" which punctuate the grain activity close to the leading edge, the development of the saltation cloud would be very considerably delayed and displaced. Removal of the spikes from the input data confirms this to be so. Thus we confirm the finding of Williams [10] that fluctuations in the near bed velocity field are critically important in initial grain motion and hence in subsequent saltation cloud development. Whether, as he postulates, the important grain disturbing features are related to the burst and sweep structures of the boundary layer we can provide no confirmation because of the shortness of our film record and the absence of continuous velocity measurement at the site of the disturbances. However, the establishment of the important role of near-bed velocity fluctuations reinforces the need for some good measurements of velocity in the grain-laden layer, both in developing and in uniform conditions.

The calculations leading to Fig. 10 have no direct wind dislodgement beyond 6 cm from the leading edge. While direct dislodgement seems to decay quite rapidly with x, there remains the possibility that vigorous features like those which generated the flurries of activity in the 1—2 cm strip, may also occur downwind, when grain transport is well-developed. Because of their significance in establishing the stable load, it is particularly important to know whether they also occur downwind, where the near-bed velocity gradient is progressively reduced by the drag exerted on moving grains.

5 Conclusions

1. For compact, quartz grains, to which our experiments were confined, most initial grain motions are in rolling, although some grains jump directly into a first trajectory.
2. Rolling grains, near the leading edge, on average gather speed until an impulsive contact with the bed provides the necessary vertical momentum for take-off.
3. Vigorous flurries of grain activity occur sufficiently near the leading edge that they must be directly wind generated. These flurries are very important in the development of general grain transport.
4. The development of the saltation layer can be modelled with reasonable success by a modified saltation cloud calculation.
5. The model will be a useful tool in verifying concepts of saltation layer development. However, it will be used with greater confidence when more is known about turbulence in the grain laden air.

Acknowledgements

The financial support of NERC has made this study possible. It has also benefited greatly from finance from NATO and the EC to support invaluable contact with colleagues at Aarhus University, Duke University and University of California, Santa Cruz.

References

[1] Bagnold, R. A.: The physics of blown sand and desert dunes. London: Methuen 1941.
[2] Mitha, S., Tran, M. Q., Werner, B. T., Haff, P. K.: The grain bed impact process in aeolian saltation. Acta Mech. **63**, 267—27 8(1986).
[3] Rumpel, D. A.: Successive aeolian saltation: studies of idealised conditions. Sedimentology **32**, 267—280 (1985).
[4] Willetts, B. B., Rice, M. A.: Collision in aeolian saltation. Acta Mech. **63**, 255—265 (1986).
[5] Anderson, R. S., Haff, P. K.: Simulation of aeolian saltation. Science **241**, 761—776 (1988).
[6] Werner, B. T.: A steady state model of windblown sand transport. J. Geol. **98**, 1—17 (1990).

[7] Rice, M. A.: Grain shape effects on aeolian sediment transport (this volume).
[8] Willetts, B. B.: Transport by wind of granular materials of different grain shapes and densities. Sedimentology **30**, 669—679 (1983).
[9] McEwan, I. K., Willetts, B. B.: Numerical model of the saltation cloud (this volume).
[10] Williams, J. J.: Aeolian entrainment thesholds in a developing boundary layer. Ph. D. Thesis, University of London (1986).

Authors' address: B. B. Willetts, M. A. Rice, and I. K. McEwan, Department of Engineering, Kings College, Aberdeen AB9 2UE, United Kingdom

Acta Mechanica (1991) [Suppl] 1: 135—144
© by Springer-Verlag 1991

Wind tunnel observations of aeolian transport rates

K. R. Rasmussen, Aarhus, and **H. E. Mikkelsen**, Tjele, Denmark

Summary. We have studied some of the causes of the discrepancies between the transport rate formulae published in the literature. The slow development of the equilibrium boundary layer over a rough bed with saltation may bias the wind gradient measurements, especially in short wind tunnels. Other discrepancies may be due to transient transport conditions caused by changes in bed texture and grain size composition. Wind profile data show that the aerodynamic roughness length $z_0 = 0.022\ u_*^2/2g$, in which u_* is the friction velocity and g is the gravity constant. This agrees well with earlier findings, and the equation could be an important tool in validating wind tunnel data. Transport rate data from the Aarhus wind tunnel differ significantly from the Lettau and Bagnold equations, but agree well with predictions from a new analytical model.

Introduction

Numerous data on the aeolian sediment transport rate and its relation to wind flow have been published during the last half century (see e.g. [1] and [2]). Most of the data originates from laboratory experiments where disturbing influences from uncontrolled variations in wind flow and sand supply can be avoided. The influence of variations in the grain size distribution and in grain shape have been included in transport dynamics. However, the agreement between relationships obtained in different experiments is surprisingly poor. We think that the following questions are important in considering the inconsistencies of the data; a) can the uptake and transport of sand grains by wind be realistically simulated in a wind tunnel? b) are the measurements of sand flux shear and stress accurate enough?

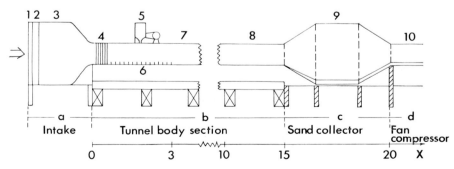

Fig. 1. Schematic view of the Aarhus wind tunnel showing the different elements. 1) Screen, 2) Honeycomb, 3) Contraction, 4) Profile modifier, 5) Sand-feed, 6) Artificial roughness elements, 7—8) Working area, 9) End-box, 10) Fan and compressor. The distance x from the upstream end of the tunnel section is measured in metres. The tunnel is 35 cm wide

In order to investigate these subjects we have in recent years conducted a series of wind tunnel experiments in the 15 m long wind tunnel at the Department of Geology, Aarhus University. The tunnel is shown in Fig. 1 and an extensive description is given in [3].

During these studies it was found that three different frequently used non-isokinetic traps catch substantially less (25—50%) than an isokinetic reference trap [4]. Hence, in the following we shall focus on problems related to wind tunnel dynamics and the assessment of wind flow parameters. In the light of our current knowledge of wind tunnel flow and trap behaviour we shall also comment on some earlier, much cited, experiments.

The rough bed and the boundary layer

Many wind tunnels used for aeolian experiments are constructed so that a uniform wind profile is created at the entry. The wall shear stress is therefore high immediately downwind from the entry but decreases quickly near the rough bed as a new boundary layer develops. (Boundary layers develop along all sides of the wind tunnel, but since the bed roughness far exceeds that of the other sides we shall neglect the influence of the sides and top on the flow). The wind speed in the boundary layer is reduced as compared to the speed U_∞ in the undisturbed free flow. As can be seen in Fig. 2 a true equilibrium between U_∞ and the surface friction speed u_{*0} is rarely, if ever, reached in any aeolian tunnel. The data points plotted in Fig. 2 are measured above a saltating sand bed at approximately 10 m downwind from the working section entry. The figure shows that for even a 100 m long tunnel the rate of change in U_∞/u_{*0} above a rough bed will still be considerable. Consequently, if we study the development of the boundary layer above a quiescent sand bed, the results may not be representative for a bed with saltation which has a much higher roughness [1], [4].

A uniform bed of immobile pebbles was made in order to create a static roughness of the same size as the roughness of a saltating sand bed. At two positions, approximately 8 m (P1) and 12 m (P2) downstream from the wind tunnel entry, the wind profiles were then measured with a pitot tube ($OD = 3$ mm) at two different wind speeds ($U1$ and $U2$). At the two positions and at two heights ($h_1 = 5$ cm, $h_2 = 8$ cm) a triaxial hot-wire probe was also used to resolve the wind vector (U, V, W) into its average ($\overline{U}, \overline{V}, \overline{W}$) and fluctuating parts (u, v, w). U is in the direction along the tunnel axis and V and W are the transverse and vertical components. Hence $\overline{V} \equiv 0$ and $\overline{W} \equiv 0$. Data were recorded at 40 Hz in 80

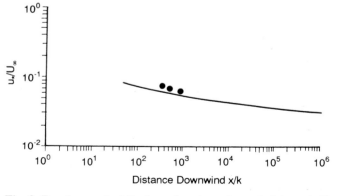

Fig. 2. Development of the boundary layer in a wind tunnel. U_∞ is the free stream velocity, x is the distance from the upwind end and k is Nikuradse's equivalent sand roughness. The curve is from [5]. The data points are measured above a sand bed with saltation in the Aarhus wind tunnel

second blocks over approximately 6 minutes (except for one run with 20 Hz sampling frequency over approximately 15 minutes).

The velocity profiles based on the pitot tube measurements are shown in Fig. 3. The value of z_0 is about 0.5 mm which is a typical value for a sand bed with saltation (see below) and at the two locations the apparent height of the boundary layer is only about 4 and 6 cm. The shape of the upwind wind profiles is similar to that reported from other wind tunnels [7]. We have not used the triaxial probe closer to the bed than about 5 cm because of the risk of flow distortion. The wind vectors calculated from the tri-axial probe deviate less than $1°$ from the tunnel axis although the probe is lined up by eye. The (small) $\overline{V}$ and $\overline{W}$ components were removed from the raw data before calculating the variances ($\overline{u^2}$, $\overline{v^2}$, $\overline{w^2}$), correlations ($\overline{uv}$, $\overline{uw}$, $\overline{vw}$), and the apparent hot-wire friction velocities (Table 1). From the

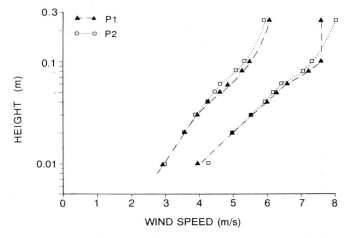

Fig. 3. Wind profiles at two locations ($P1$ and $P2$) measured above a rough static bed. $P1$ is immediately upwind of the working area ($x \approx 8$ m), while $P2$ is well within this ($x \approx 12$ m). Data are given for two overall air speeds $U1$ and $U2$, and the corresponding friction speeds derived from the profile at $P2$ are $u_{*U1} = 0.38$ m/s and $u_{*U2} = 0.52$ m/s

Table 1. Variances, correlations and normalized standard deviations based on hot-wire measurements at two positions in the wind tunnel, $P1 = 8$ m and $P2 = 12$ m from the entry. Date were obtained at two height, $h_1 = 5$ cm and $h_2 = 8$ cm above the bed and for two overall wind speeds, $U1$ and $U2$

			$\overline{u^2}$	$\overline{uv}$	$\overline{uw}$	$\overline{v^2}$	$\overline{vw}$	$\overline{w^2}$	U	σ_u/u_*	σ_v/u_*	σ_w/u_*	u_*
$P1$	h_1	$U1$	0.421	0.005	-0.111	0.289	0.024	0.255	4.633	1.70	1.40	1.33	0.333
40 Hz		$U2$	0.754	-0.003	-0.182	0.601	0.068	0.501	5.951	1.66	1.48	1.35	0.426
	h_2	$U1$	0.383	0.001	0.003	0.240	0.034	0.295	4.956	—	—	—	—
		$U2$	0.550	0.023	-0.125	0.467	0.055	0.381	6.899	1.41	1.74	1.50	0.354
$P2$	h_1	$U1$	0.425	0.008	-0.104	0.273	0.022	0.233	4.322	1.71	1.37	1.26	0.322
40 Hz		$U2$	0.770	0.020	-0.186	0.553	0.047	0.468	5.974	1.67	1.41	1.23	0.431
	h_2	$U1$	0.310	0.007	-0.082	0.205	0.022	0.185	4.948	1.46	1.18	1.12	0.286
		$U2$	0.549	0.024	-0.141	0.414	0.029	0.369	6.859	1.41	1.23	1.16	0.375
$P2$	h_1	$U2$	0.732	0.023	-0.168	0.560	0.046	0.461	5.941	1.63	1.43	1.30	0.410
20 Hz													

apparent friction velocities measured at $P2$ we have extrapolated the surface value u_{*0} during the runs. For $U1$ we find $u_{*0} = 0.38$ m/s while at the higher speed $U2$ $u_{*0} = 0.52$ m/s. These values are used in calculating the normalized standard vediations (σ_i/u_*) for the three velocity components (see Table 1).

The friction velocities based on the lower part of the velocity profiles at $P2$ agree very well with the hot-wire results. Further, in the full-scale boundary layer at neutral stability the canonical values of the normalized standard deviations are $\sigma_u/u_* = 2.2$, $\sigma_v/u_* = 1.7$ and $\sigma_w/u_* = 1.3$. In our wind tunnel all three normalized standard deviations are generally lower than the canonical values, probably due to the suppression of large scale fluctuations. For all three components the standard deviations are closest to the full scale values near the bottom where the boundary layer is best developed. Further, the values seem to increase with increasing U_∞, possibly indicating a dependence on Reynold's number. Finally, there is virtually no correlation between the u and w components for the low speed measurement at 8 cm at the upstream position. Consequently, this reading must be entirely within the free stream. The turbulence measurements support that the fully developed boundary layer flow does not reach much higher than 5 cm, especially not at the upwind position.

Steady state transport in a wind tunnel

Much of the current progress in understanding the saltation process has been based on work with uniform grain populations ([8], [9], [10]). However, it is important to ask whether the interaction of the transport processes with a natural sand bed composed of a range of size classes in itself prevents steady state transport from developing. It is well known that saltation starts through aerodynamic lift on the bed particles when the surface shear stress exceeds the fluid threshold [11]. Since the fluid threshold is a moderate function of particle diameter [12] the start of saltation in a natural sediment may be a selective process. During steady state transport the volume of moving sand is primarily restricted to the sand contained in the ripples [13]. When these move downwind all grains are gradually exposed to the air flow and since the average speed of creeping grains is less than that of saltating grains the latter may progressively be winnowed from the bed, leaving the creeping grains as a residual "armouring". This subsequently influences saltation mechanics: it leads to higher and longer saltation trajectories, thereby causing a change in the ripple shape and in transport conditions. Conditions in nature may be different due to a much larger upwind fetch from which sand of all grain sizes may be supplied by advection.

In order to test the steady state concept a series of measurements were performed with a natural wind-blown sand placed in a 3 cm thick layer in the wind tunnel. The sand was well sorted, with a median grain size near 220 μm [4] and a threshold friction velocity $u_{*t} \approx 0.25$ m/s. At the start of each experiment the bed material was mixed well and scraped flat before the fan was turned on for 6—10 minutes to develop a stable ripple pattern. We have measured the transport-rate profile ($\phi(z)$) in consecutive intervals during one to two hour long runs. For three heights we calculated the relative transport rates $R_\phi = \phi(z)/\phi(z)_0$,where $\phi(z)_0$ is the value during the first 10 minutes, and plotted these in Fig. 4. R_ϕ is almost constant during the first 30 minutes — we shall refer to this as *the initial transport rate* — but subsequently it decreases markedly. The bed texture gradually starts to coarsen shortly after the start of an experiment. This is accompanied by a change in the ripple shape which becomes longer and more sharp-crested. Data from an earlier experiment can be used to illustrate this. There the ripple pattern was extensively mapped for

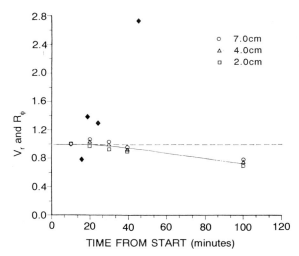

Fig. 4. The relative transport rate R_ϕ as function of elapsed time and at three heights (∘ , □ , △). Values are normalised with the transport rate measured during the first ten-minute period, and are shown at the end of each sampling period. The dimensionless ripple volume V_r is also presented (♦), but data for this are only measured during a 45 minute run

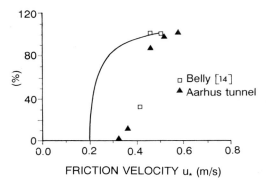

Fig. 5. The ratio (in %) between transport rate respectively without sand feeding and with sand feeding in the tunnel. Data is from for the Aarhus wind tunnel and from [14]. For the Aarhus data we also show the weight percentage of the sand that is above u_{*t} at a given u_*-value (solid line)

45 minutes and in Fig. 4 is plotted the relative change in ripple volume $V_r = \lambda h/\lambda_0 h_0$ where λ is the ripple length, h is the ripple height and $_0$ denotes the initial values. From Fig. 4 it is evident that transport conditions remain stationary for only about 30 minutes. Because of this the duration of later runs was restricted to 30 minutes and as a result the systematic error in the transport rate data at a given height and for a fixed wind speed is less than 5%.

The influence of the sand feed can be seen from a comparison of pairs of otherwise identical runs but where the sand feed was turned on in one run and off in the other (Fig. 5). For u_* below about 0.45 m/s the transport rate is consistently smaller with the sand feed turned off. Using the relation between threshold friction speed u_{*t} and grain size d given in [12] we calculated the relative percentage of the bed material (by mass) for which, for a given value of u_{*t}, $u_* > u_{*t}$. From Fig. 5 it appears that the sand feed influences the sand transport until the friction velocity is above its threshold value for almost the entire grain population. At lower speeds the momentum necessary to initiate the chain reaction that starts saltation is presumably provided via grains falling from the sand feed. Corroborating wind tunnel observations were made using sand with a modal value of 440 μm [14], as well as soil [15]. The values found in [14] are also plotted in Fig. 5.

The wind profile in saltating flow

The velocity data (Fig. 6) were recorded at heights between 20 and 80 mm at the downstream end of the experimental section. The profiles follow the logarithmic wind law almost perfectly

$$u(z) = \frac{u_*}{\varkappa} \ln \frac{z}{z_0} \tag{1}$$

where $\varkappa$ is von Karman's constant. From the extrapolation of the wind profiles to zero velocity we can see that z_0 increases monotonically with increasing wind speed. This is due to momentum being extracted from the air flow at a progressive rate. From Fig. 6 we can also see that the partitioning of the shear stress between the air flow and the saltating grains [11] does not influence the velocity profiles at heights between 20 and 80 mm. This is in accordance with numerical predictions in [16] since our friction velocities are low or moderate and most of the transport ($\approx 75\%$) takes place below a height of 20 mm [17]. In [1] it is proposed that

$$z_0 = C_0 \frac{u_*{}^2}{2g} \tag{2}$$

with $C_0 = 0.021$; g is the acceleration of gravity. We have plotted z_0 as a function of $u_*{}^2/2g$ (Fig. 7) and find $C_0 = 0.022$. From Eq. 2 we see that for a saltating sand bed z_0 will be of the order of 10^{-4}—10^{3-} m.

We assume that Eq. 1 is valid throughout the saltation layer and combine Eq. 1 and Eq. 2. For a given friction velocity $u_* > u_{*t}$ we have

$$\zeta_0 = \frac{C_0}{2g} \exp \frac{2(u_* \ln u_* - u_{*t} \ln u_{*t})}{u_* - u_{*t}} \tag{3}$$

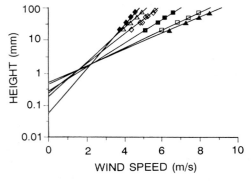

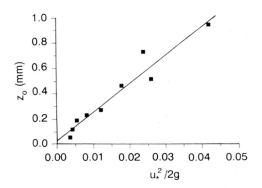

Fig. 6. Saltation layer wind profiles measured for six different free stream velocities

Fig. 7. Roughness length z_0 shown as function of $u_*{}^2/2g$. The slope of the regression line (C_0) is 0.022 when both variables are expressed in the same units

where ζ_0 is the height at which the log-linear wind profile intersects the wind profile of the threshold friction velocity. Using a typical value of $u_{*t} = 0.25$ m/s the value of ζ_0 will vary from about 0.5 mm to 1.5 mm for u_* in the range 0.3—0.7 m/s. With the assumptions given above, our findings do not strictly support the existence of a focal point [11].

Aeolian transport rate

The sand flux $Q = \left(\int \phi \, dz \right)$ was calculated from data measured with the isokinetic trap. According to [1] the energy that the grains extract from the air per unit time is $\varrho u_*^3/g$ and this is used to normalize the sand flux data and estimate the dimensionless sand transport flux $\Phi = Qg/\varrho u_*^3$. In Fig. 8 Φ is plotted as a function of u_*/u_{*t}. For low friction velocities Φ increases gradually from the threshold value. Thereafter a maximum is reached for $u_*/u_{*t} \approx 2$ after wihch Φ drops slowly.

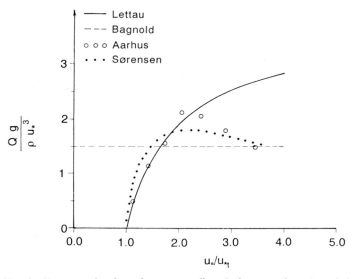

Fig. 8. The normalised sand transport flux Φ shown as function of the dimensionless friction velocity u_*/u_{*t}. Curves corresponding to the normalized "Bagnold" (Φ_B) and "Lettau" (Φ_L) transport rate formulae are also shown. Finally, the predicted normalized flux from an analytical model by Sørensen [19] (Φ_S) is shown

The "Bagnold equation" [11] assumes that Q is proportional to u_*^3 and depends on the grain size. We write it in the dimensionless form

$$\phi_B = \frac{Qg}{\varrho u_*^3} = C_B \left(\frac{D_p}{D_{p0}} \right)^{1/2} \tag{4}$$

where D_p is the median grain diameter and D_{p0} is a scaling diameter equal to 250 μm. The value of C_B given by Bagnold is 1.5—2.8, but other authors have found very different values. When plotted as a function of u_*/u_{*t} Φ_B in Eq. 4 is a horizontal line (Fig. 8), and $C_B \approx 1.5$ for a best fit to our data. However, the structure of our data differs considerably from the Bagnold equation.

The initial jump of Φ_B at u_{*t} as exhibited by Eq. 4 is avoided in the "Lettau equation" [18]. When normalized as above this equation has the form

$$\phi_L = \frac{Qg}{\varrho u_*{}^3} = C_L \left(1 - \frac{u_{*t}}{u_*}\right) \tag{5}$$

and a value of $C_L = 5.5$ was suggested [18]. A reasonable fit to our data is obtained with $C_L = 3.8$ (Fig. 8). The curve describes the data well for $u_* < 2.5 u_{*t}$, but it overestimates the transport rate for larger values of u_*.

Recently, an analytical model for saltation [19] was formulated. From this the normalized transport rate Φ_S can be written in the form

$$\phi_S = \frac{Qg}{\varrho u_*{}^3} = c_1 \left(1 - \frac{u_{*c}}{u_*}\right) \left(1 + \frac{7.6 u_{*c} + 205}{u_*}\right) \tag{6}$$

where u_{*c} is 0.25 m/s and $c_1 = 0.0014$. The best fit to our data is obtained for $c_1 = 0.0004$ for which Eq. 6 is also shown in Fig. 8. It is interesting to see that Eq. 6 fits the structure of our data much better than the earlier equations.

We now wish to comment on some earlier studies of sand transport and in particular point out the difficulties in finding reliable data for comparison with the current results. In [20], sand transport was studied in a 0.8 m high and 5 cm wide wind tunnel with a 2.5 m long working section. Presumably, wall effects did influence wind measurements, and it is not clear how high the boundary layer was. Velocities measured at height between 10 cm and 30 cm were used to derive a relation between u_* and the wind speed at 30 cm. From a recalculation of the data we find a relationship between z_0 and $u_*{}^2$ that differs strongly from our Eq. 2. Probably the intermediate and high friction velocities presented in [20] do not fully reflect the flow properties at the bed and the formula will overestimate the transport rate.

Several authors have not measured friction velocity directly, but inferred it from extrapolation based on a measured velocity at some height in the tunnel and a presumed focal point, see [14] and [21]. Although the wind tunnel used in [14] is long, inspection of the wind data suggests that the boundary layer perhaps was not fully developed to the height of 1 ft where the reference velocity was measured. Another, probably more important objection to this method is that any changes in bed texture, and thus z_0, will change the relation between the reference speed and u_*.

Turning to results obtained in [21], we have no reason to mistrust the wind data, but the transport rates appear to have been obtained with a non-isokinetic trap. Moreover, no sand feed seems to have been used to initiate bed transport. The latter deficiency may not be very serious, since the smallest friction velocity is about 0.4 m/s, but poor trap efficiency could lead to an underestimation of the transport rate. In fact, an increase of the transport rates would bring the results closer to our own data.

Discussion and conclusions

Flow over a rough surface in a wind tunnel has been studied by means of velocity profile and turbulence data. The equilibrium boundary layer (with respect to first and second order moments) is very shallow. After an upwind fetch of 8 m it is only about 5 cm high. Therefore, when deriving the friction velocity from profile data it is necessary to use a long tunnel in order to form a sufficiently thick layer for precise wind measurements. Furthermore, if no proper boundary layer is developed, the saltation dynamics will agree poorly with conditions in nature.

Several factors may influence transport dynamics in a wind tunnel and cause transient conditions. In our 15 m long tunnel we find that the sand transport is quasi-stationary for approximately 30 minutes. Thereafter bed texture and sand transport rate changes markedly. Further, when the friction velocity is below the threshold value for the majority of the grains a sand feed is necessary to obtain a stationary transport rate. We think it is important to clarify in what way wind tunnel length affects the development of a stationary sand transport.

We have obtained new data on how the transport rate relates to shear stress in a wind tunnel. Our data confirm that $z_0 = C_0 u_*^2/2g$ with $C_0 \approx 0.022$. If additional experiments can verify this relationship, or provide indications as to how C_0 varies with grain size, then the reliability of wind data can be checked. We have compared our transport rate data to the Bagnold and the Lettau formulae and a completely new equation [19], and the overall structure of the data agrees best with the last. There are many factors that may bias our measurements, and with the present limited number of data we do not wish to be too conclusive. However, there is now experimental as well as theoretical evidence for a structure in the dimensionless transport flux that is different from recent numerical calculations [8] as well as the "classic" transport rate equations.

Acknowledgements

This work has been done with support from the Danish Natural Science Research Council, the Danish Technical Research Council, the NATO Science Fellowhip Programme and from Aarhus Universitets Forskningsfond.

References

[1] Owen, P. R.: Saltation of uniform grains in air. J. Fluid Mech. 20, 225—242 (1964).

[2] Greeley, R., Iversen, J. D.: Wind as a geological process on Earth, Mars, Venus and Titan. London: Cambridge University Press, pp. 333, 1985.

[3] Rasmussen, K. R., Mikkelsen, H. M.: Development of a boundary layer wind tunnel for aeolian studies. Geoskrift No. 27. Geologisk Institut, Aarhus Universitet, 1988.

[4] Rasmussen, K. R., Mikkelsen, H. M.: Aeolian transport in a boundary layer wind tunnel. Geoskrift No. 28. Geologisk Institut, Aarhus Universitet, 1988.

[5] Schlichting, H.: Boundary layer theory, 6th ed. New York: McGraw-Hill, pp. 653, 1968.

[6] Gerity, K. M.: Problems with determination of u_* from wind-velocity profiles measured in experiments with saltation. In: Proceedings of International Workshop on the Physics of Blown Sand. Memoirs No. 8. Department of Theoretical Statistics, Aarhus University 1985.

[7] White, B.: Two-phase measurements of saltating turbulent boundary layer flow. Int. S. Multiphase Flow 8, 459—473 (1982).

[8] Anderson, R. S., Haff, P. K.: Simulation of eolian saltation. Science 241, 820—823 (1987).

[9] Ungar, J. E., Haff, P. K.: Steady-state saltation in air. Sedimentology 34, 289—299 (1987).

[10] Werner, B. T.: A steady-state model of wind blown sand transport. J. Geology 98 (No. 1), 1—17 (1990).

[11] Bagnold, R. A.: The physics of blown sand and desert dunes. London: Methuen, pp. 265, 1941.

[12] Iversen, J. D., White, B. R.: Saltation threshold on Earth, Mars, and Venus. Sedimentology 29, 111—119 (1982).

[13] Barndorff-Nielsen, O. E., Jensen, J. L., Nielsen, H. L., Rasmussen, K. R., Sørensen, M. S.: Wind tunnel tracer studies of grain progress. In: Proceedings of International Workshop on the Physics of Blown Sand. Memoirs No. 8. Department of Theoretical Statistics, Aarhus University 1985.

[14] Belly, P. Y.: Sand movement by wind. Technical Memorandum No. 1. U.S. Army Coastal Eng. Res. Center, Washington D. C., pp. 80, 1964.

[15] Chepil, W. S.: Properties of soil which influence soil erosion I. The governing principle of surface roughness. Soil Sci. 69, 149—162 (1950).

[16] Rasmussen, K. R., Sørensen, M., Willetts, B. B.: Measurement of saltation and wind strength on beaches. In: Proceedings of International Workshop on the Physics of Blown Sand. Memoirs No. 8. Department of Theoretical Statistics, Aarhus University 1985.

[17] Rasmussen, K. R., Mikkelsen, H. M.: The transport rate profile and the efficiency of sand traps. Sedimentology (in press).

[18] Lettau, K., Lettau, H. H.: Experimental and micro-meteorological field studies of dune migration. In: (Lettau, H., Lettau, K., eds.) Exploring the world's driest climate. Institute for Environmental Studies, University of Wisconsin-Madison, pp. 264, 1978.

[19] Sørensen, M.: An analytic model of wind-blown sand transport (this volume).

[20] Kawamura, R.: Study on sand movement by wind. Reports of Physical Sciences Research Institute of Tokyo University, Vol. 5, No. 3—4, 1951, p. 95—112. [Translated from Japanese by National Aeronautic and Space Administration (NASA), Washington D. C. 1972].

[21] Zingg, A. W.: Wind tunnel studies of movement of sedimentary material. Proceedings of the 5th Hydraulic Conference Bulletin 34, 111—134 (1953).

Authors' addresses: K. R. Rasmussen, Institute of Geology, University of Aarhus, DK-8000 Aarhus C, and H. E. Mikkelsen, Department of Agrometeorology, Research Center Foulum, DK-8830 Tjele, Denmark.

Acta Mechanica (1991) [Suppl] 1: 145—157
© by Springer-Verlag 1991

An experimental study of Froude number effect on wind-tunnel saltation

B. R. White and **H. Mounla**, Davis, CA, U.S.A.

Summary. The simulation of the natural process of saltation in a wind tunnel is considered. The pioneering interactive boundary-layer analysis of Owen and Gillette [11] concluded that an independence Froude number criterion did apply to the problem and they estimate, based on wind-tunnel Froude number data ranging from 35 to 80, an independence Froude number value of about 20 for saltating flows to be free of facility constraints imposed on the saltation. The present experimental flows had Froude numbers ranging from about 6 to 1000. Analysis of friction speed variation as a function of downstream position suggests a more conservative critical Froude number value of 10 be used. Also, there appears to be an additional requirement, for most of our data, that tunnel's downstream length-to-heigt ratio be greater than 5. Therefore, a maximum Froude number of 10 and minimum tunnel length-to-height ratio of 5 is suggested to insure accurate saltation tunnel simulation.

1 Introduction

This study expands upon the work by Owen and Gillette [11] and Owen [10], which sets operation limits on wind tunnels, and analyzes variation of the friction speed as a function of the downstream position for various Froude numbers in the presence of a saltating particle bed. The dimensions of the wind tunnel are important factors in determining if a choked saltation flow will occur, resulting in a blockage of the airflow. Under choking conditions, particle flux and velocity profiles are altered by the constraints placed upon the flow by the wind tunnel. Thus, inaccurate modeling will occur. According to Owen and Gillette, a wind tunnel operated at a speed which has a value of the Froude number less than 20, should be free from this effect, resulting in small variations of u_* as a function of the downstream position.

The wind-tunnel experiments, carried out by Owen and Gillette, were for a relatively small tunnel (19.5 cm high) where the determinate influence of Froude number was readily noticeable. Their experimentally determined values of Froude number were all in excess of 35 or almost double their recommended maximum value of 20 for flows to be free of Froude number effects. In observing Fig. 4 of their paper, it may be noticed that few, if any, of the saltating flows achieve a constant value of friction speed u_* with downstream distance. This result is to be expected since the Froude number independence principle (i.e., Froude number values less than 20) was not met in their experiments.

One of the main objectives of the present study was to carry out windtunnel experiments in which the Froude number criterion was met and then determine with a relatively high degree of accuracy the variation of u_* as a function of both downstream distance and Froude number. If the Froude number criterion is met and the tunnel is free of entrance effects, then u_* should approach or obtain a constant value. Accordingly, saltation experiments have been conducted in a UC Davis boundary-layer wind tunnel and in NASA's low-pressure MARSWIT wind tunnel that have produced a range in Froude number from about 6 to 1000. Thus, we were able to test the hypothesis of the Froude number independence principle over a wide range of Froude numbers.

Lastly, many mathematical formulations predicting mass transport rates (the mass of particles, M, passing through a finite width, w, during a time interval T) are found in the literature (a nice summary of these formulae is found in Greeley and Iversen [3]). Nonetheless, a limited number of these expressions are more favored than the others, as they are considered more accurate. The reason for the discrepancy in wind-tunnel measurements of transport rates is, in part, attributed to the collector's efficiency, which is a function of the wind tunnel as well as aerodynamic features of the collector's shape. Therefore, it is of importance to determine an efficient particle collector for any tunnel. The present study summarizes the results of such an investigation; i.e., the optimum particle collector was determined experimentally for the UC Davis boundary layer wind tunnel. Additionally, mass flux as a function of height as well as total mass fluxes were measured. Simultaneous measurements of the mass flux rate along with velocity profiles at the same location were also made.

2 Wind-tunnel facility and measurement techniques

The majority of the saltation experiments were performed in a boundary-layer wind tunnel located at UC Davis shown schematically in Fig. 1. The tunnel was designed to simulate particle flow in saltation, and it consisted of four major sections: (i) entrance, (ii) flow development, (iii) test, and (iv) diffuser. The overall length of the tunnel was 13 m.

The entrance section consisted of a 0.48 m long contraction area having a ratio of 5:1 equipped with honeycomb flow-straighteners to reduce the freestream turbulence level. The 7.32 m flow development section had diverging walls and was composed of three individual sections which were each 2.44 m (8 feet) long made of plywood. The Plexiglas test

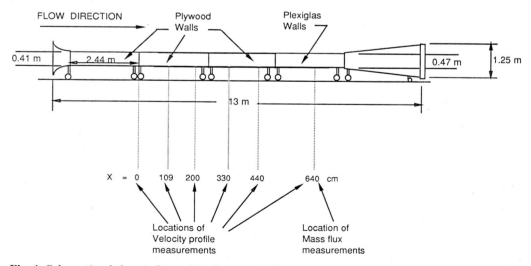

Fig. 1. Schematic of the windtunnel facility at the University of California, Davis

section was 2.44 m in streamwise length and 0.47 m high by 0.8 m wide in cross section. The diffuser, 2.8 m in length, had an expansion area ratio of 13:1 which provided a continuous transition from the rectangular cross section of the test section to a circular cross section for the fan.

The tunnel was equipped with a 3 hp, variable speed *DC* motor that rotated an 8-blade fan. An *AC/DC* power converter supplied power to the motor, and had controllable air speed inside the tunnel of up to 15 m/s.

3 Mean velocity profiles in saltation

A series of experiments, in the presence of saltating particles, were carried out to measure the mean velocity profiles and determine the effect of Froude number on the upstream of the bed. The Froude number was defined as $U_1^2/(gH)$, where U_1 is the inviscid uniform velocity upstream of the bed, g is the gravitational acceleration and H is the tunnel height.

A 1 cm deep smooth bed of walnut shells was placed on top of the nonerodible walnut shell surface. The bed started 2.4 m downstream from the entrance and extended about 7.5 m downstream to the end of the test section.

A Pitot-static tube, mounted on a transversing mechanism, moved vertically from the floor of the tunnel upward to a height of 31.2 cm to determine the mean velocity probe.

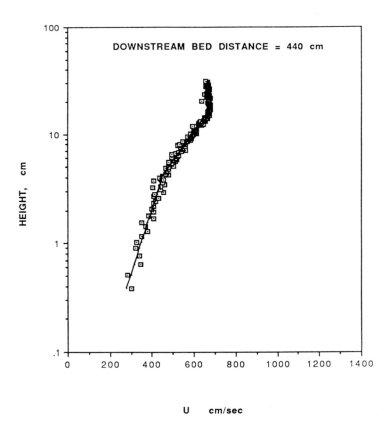

Fig. 2. Typical measured velocity data (240 values) as a function of logarithmic height. The two solid lines represents a linear regression fit to the data

The Pitot-static tube was connected to the Barocel pressure system which in turn was electronically connected to an A/D system controlled by an IBM AT PC. The data acquisition system operated at a rate of 1000 Hz and the program averaged 500 pressure readings acquired in 0.5 seconds to yield one data point of the velocity profile. Each profile consisted of 240 time-averaged individual points. A single profile required two minutes to be measured. Figure 2 displays a typical profile acquired at a downstream distance of 6.84 m from the entrance with a freestream speed of 6.46 m/s.

The aerodynamic roughness height of the loose bed of material was determined by measuring several velocity profiles, each at a different freestream speed, all below threshold condition. The value of the surface roughness was found to be 25 microns.

4 Determination of friction speed in saltation

The friction speed, u_*, in the presence of saltating particles, could not be determined by using the simple slope method, because of the ambiguity in the velocity profile. However, when the wake region is accounted for in much the same manner as Coles' law-of-the-wake [2] equation for smooth-wall turbulent boundary layer case a reasonable estimate of u_* may be obtained.

Therefore, the rough-wall case, with saltation occuring, may be represented by

$$\frac{u}{u_*} = \frac{1}{k} \ln_e (y/y_0) + \frac{\Pi}{k} \left(1 - \cos (\pi y/\delta)\right)$$

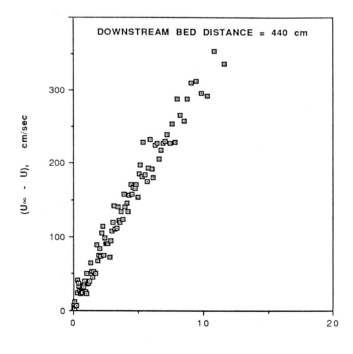

Fig. 3. The defected velocity as a function of F. The mean slope of the data is equal to the friction speed

where k is von Karman's constant, equal 0.418; and Π is Coles' wake parameter which is a function of pressure and momentum-deficit Reynolds number; here it is assumed equal to 0.55 (White [13]).

Applying this relation at the edge of the boundary layer and at an arbitrary location within the boundary layer yields (subtracting one from the other according to Mounla [8]),

$$\frac{u_\infty - u}{u_*} = \frac{-1}{k} \ln (y/\delta) + 1.32 \left(1 + \cos (\pi y/\delta)\right) = F$$

or

$$u_\infty - u = u_* F$$

Note, the form of the rough-wall velocity defect, $u_\infty - u$, is essentially the same as the smooth-wall case, since the additive constant in both cases disappears when the difference of u_∞ from u is taken.

Figure 3 presents the difference between the freestream velocity and the instantaneous velocity within the boundary layer as a function of F. The slope of this plot corresponds to the friction speed. Thus, the friction speed may be determined from the velocity profiles measured during saltation. This calculation is needed to determine the constancy of u_* with downstream position and Froude number.

5 Composite mean velocity profiles in saltation

Composite mean velocity profiles were measured during saltation at different locations along the tunnel to determine the friction speed variation as a function of downstream location. To insure uniform test conditions, at the end of each previous run, the particle bed was reconditioned to its original 7.5 m length and position, by replacing particles over the eroded portion of the surface and again smoothing it. The bed length had to be maintained constant, since the variation of the friction speed as function of the downstream location was to be investigated.

The mean velocity profiles (logarithmic height versus speed), in the presence of saltating particles, follow a straight line from the surface and then experience a deflection point at a certain height to follow another straight line to the edge of the boundary layer. Figure 4 displays the composite profiles (A, B, C, and D on each figure) at downstream bed positions of 1.1, 2.0, 3.3, 4.4, and 6.4 meters from the beginning of the bed. The lines plotted are linear regressions determined from the actual velocity profile data (such as the linear regression lines displayed in Fig. 2). Great care was taken to match the two segments of each profile at their interface as well as match the local freestream speed, u_∞, above the boundary layer. The inviscid upstream velocity for each of the profiles A, B, C and D was held constant for each downstream profile measurement; note however, locally, the freestream values of the curves A, B, C and D changed as a function of downstream position. Table 1 presents this information. The value of the threshold friction speed for the walnut shell, at a downstream distance of 640 cm was found to be 22.5 cm/s.

Over a non-saltating rough surface, the mean velocity profiles for different speeds as functions of logarithmic height theoretically coalesce at a focus point (at zero speed) to define the roughness height y_0, as given in the relation,

$$\frac{u}{u_*} = \frac{1}{k} \ln_e (y/y_0)$$

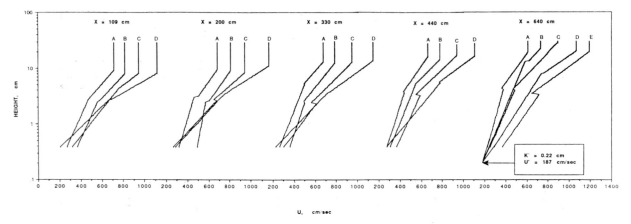

Fig. 4. Composite velocity profiles as a function of logarithmic height. The solid lines are linear regressions of the individually measured profiles. The individual values of friction speed are found in Table 1, except for curve E at downstream position 640 cm, which had a friction speed of 70 cm/s

Table 1. Calculated values of friction speed as a function of downstream position as determined from the experimental measured mean velocity profiles

Curve	Froude number	$U1$ cm/sec	$X = 109$ cm $X/H = 2.56$ u_*, cm/sec	$X = 200$ cm $X/H = 4.58$ u_*, cm/sec	$X = 330$ cm $X/H = 7.32$ u_*, cm/sec	$X = 440$ cm $X/H = 9.49$ u_*, cm/sec	$X = 640$ cm $X/H = 13.6$ u_*, cm/sec	Owen and Gillette (1985) u_*, cm/sec
A	10.4	646	43.9	37.5	39.1	34.3	36.4	33.4
B	13.8	748	47.7	34.8	39.6	43.9	38.6	40.7
C	18.6	866	58.4	56.8	51.4	48.8	48.8	50.0
D	27.6	1 060	90.0	79.3	76.1	66.4	64.8	66.6

Note: X is the distance from the leading edge of the tunnel. H is the inside tunnel height (47 cm). $U1$ was the inviscid upstream speed. The Froude defined as $U_1^2/(Hg)$

However, for a saltating flow, this focus point is shifted to finite speed, u', and a different height, k', called the roughness height in saltation (Bagnold [1]). The speed u' is generally associated with the perceived speed at which the surface particle appears to move.

In order for this phenomenon to be measured, the saltation process must have reached an equilibrium condition, i.e., constancy of u_* as a function of downstream position. The composite velocity profile of Fig. 4 displays this $u' - k'$ phenomenon. At a downstream distance of 640 cm from the beginning of the bed, an easily observed u' of 187 cm/s and k' of 0.22 cm are seen. These values of k' and u' compared reasonably well with the data published by Bagnold ($u' = 250$ cm/sec and $k' = 0.30$ cm), and Zingg [17] who reported $k' = 10\,d$ and $u' = 895\,d$, where d is the mean particles diameter in mm and k' and u' are expressed in cm, or for the present case $u' = 224$ cm/s and $k' = 0.25$ cm. Bagnold's mean particle diameter was about 270 microns, while the present experiments diameter was 250 microns.

6 Froude number effect on saltation

For each constant Froude number test, the Froude number was based on the inviscid velocity upstream of the bed, i.e., $Fr = U_1^2/(gH)$. The friction speed was determined as a function of the normalized downstream location for each Froude number case tested and is shown in Fig. 5. For a given Froude number, u_* was observed to approach a constant value as the x/H parameter was increased.

A comparison of Fig. 5 with Fig. 4 of Owen and Gillette [11] is appropriate. Owen and Gillette's case considered a smooth surface upwind of the bed thus providing an initial c_f' (or $2u_*^2/U_1^2$) that was lower than the downstream values; however, in the present experiments the upstream value of c_f' was larger than the downwind surface, i.e., the current experiment had no smooth upwind surface. These different set of inlet conditions explain why the Owen and Gillette curves asymptotically increase with increasing downstream distance while the present data decrease with increasing downstream distance.

The current experiments have Froude numbers that range from 10.4 to 27.6, some of which are below the minimum independence number of 20 set by Owen and Gillette. The trends of Fig. 5 suggest that the u_* curves are approaching a constancy value and the smaller the value of Froude number the more rapidly the constancy of u_* is approached. Also shown on Fig. 5 are the asymptotic values of u_* as calculated from the Owen and Gillette theory (see Table 1). The agreement between theory and experiment is remarkable. However, even in the the cases where the Froude number was less than 20, it took a minimum x/H

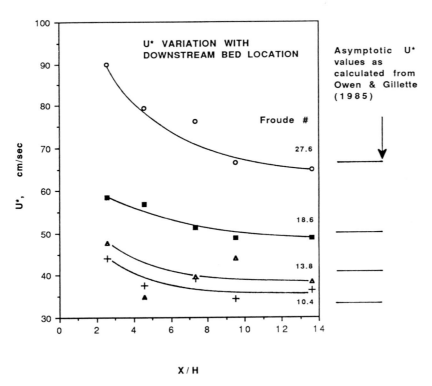

Fig. 5. The friction speed as a function of nondimensional downstream length x/H. Also shown, to the right of the figure, are the predicted values of u_* (Owen and Gillette [11]) as downstream distance becomes large

distance of at least 5 for the u_*-curves to become relatively constant with further changes in downstream position.

The air flow that enters the wind tunnel is initially almost uniform. There is a minimum entrance length required for the velocity defected layer to resemble a boundary-layer profile. Spires and other boundary layer tripping devices can shorten this entrance length; nevertheless, some minimum length still will be required to obtain a boundary layer velocity profile. The observed minimum x/H value of 5 in the current data illustrates, experimentally, the requirement of this minimum entrance length. For downstream locations less than $5H$, the friction speed has not reached its steady-state value. Consequently, the saltation process has not achieved equilibrium.

Perhaps a more refined measure, instead of x/H, would be x/δ, where δ is the boundary layer height. Generally, an absolute minimum of 25δ is required to produce an equilibrium flow (non-saltating) in terms of mean boundary-layer characteristics, such as the velocity profile (White [13]). Values of x/δ from 50 to 500 are further required to insure an equilibrium flow exists in the more subtle indictors such as turbulence intensities, development of inertial subrange and energy spectra. Fortunately, these more stringent guidelines do not apply directly to saltating flows since the particle motion severely modifies the boundary layer. Nonetheless, 25δ as a minimum entrance length requirement should be adhered to in order to establish equilibrium (constant friction speed) saltation flows. This assumes the Froude number is less than the critical value.

Some previously unpublished (Leach [7] and Iversen [5]) mass transport data from the MARSWIT low-pressure wind tunnel (Greeley et al. [4]), located at NASA Ames Research Center, Moffett Field, California, is also available for consideration. Figure 6 displays a

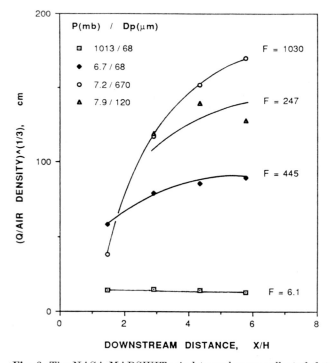

Fig. 6. The NASA MARSWIT wind-tunnel mass collected data as a function nondimensional downstream distance x/H. The Froude numbers vary from 6.1 to 1 030. P is the ambient pressure in millibars while Dp is the mean particle diameter in microns, after Iversen [5]

normalized mass collection parameter, $(M/\varrho)^{1/3}$, which is proportional to the friction speed, as a function of downstream position for the four Froude number cases tested. Notice, the ambient pressure varies from 1013 mb (atmospheric pressure) to 6.7 mb (low pressure). The low-pressure saltation experiments have relatively higher freestream wind speeds which result in much larger values of Froude number (to order velocity squared). If u_* is constant with downstream distance, then the mass collection parameter, $(M/\varrho)^{1/3}$, should also be constant with downstream distance. The observed trends in the data suggest that for large values of Froude number (greater than 100) the friction speed is not constant. Owen and Gillette [11] conclude from their analysis that when the Froude number is greater than 100, saltating particles will collide with the tunnel ceiling which has been observed in the MARSWIT tunnel when operated at low-pressure conditions (see Fig. 5; White [14]). For the atmospheric-pressure case (Froude number equal 6.1) u_* is constant as a function of downstream distance.

In summary, the current data trends suggest there is strong support for the idea put forth by Owen and Gillette of a critical Froude number required for independence of wind-tunnel effects. However, there is a lack of experimental data available to accurately assess the value of this independence Froude number. It seems as though the value of 20 may be liberal and a more conservative value of 10 would further insure a wind tunnel to be free of Froude number effects or constraints. Separately, a minimum entrance length of 25δ is required.

7 Saltating mass as a function of height

The distribution of mass flux as function of height also was measured. The particle collection system consisted of 25 individual 2 cm high, stackable Plexiglas collectors as described in White [14]. Each collector had approximately 2 cm² frontal cross-sectional area and a wire mesh at the back wall which allowed air and mean-sized particles of 40 microns diameter or less to pass through it. The air flow through the rear wall area of the collectors prevented separation from occurring off the sharp-edged intersection of the side wall with the rear wall. Flow visualization studies showed particles to be efficiently trapped. Once the collectors were positioned inside the tunnel and the freestream wind speed achieved, an automated collection system commenced. The length of the collection time for calculation of flux rates was performed by a stop watch with an accuracy of 0.01 seconds. This introduced a negligible uncertainty since collections times were typically greater than 30 seconds. After the particles were caught in the individual traps, the material of each trap was weighed by a Metler scale (accuracy of 0.1 g).

The mass flux as a function of height was measured at a downstream distance of 640 cm; simultaneous measurements of velocity profiles also were performed at the same location, see Fig. 4, $x = 640$ cm. The logarithm of the mass flux was plotted as a function of the height of each collector as shown in Fig. 7. This plot should be expected to trace a straight line, as previous work has shown an exponential dependence of mass distribution with height (White, [14] and others). At higher speeds, the appearance of a linear logarithmic relation is better than it is at lower speeds. This could be attributed to the accuracy of the scale used to measure the mass collected for a particular run. Also, at lower speeds, the uncertainty in the measurements becomes large since the collected mass is small.

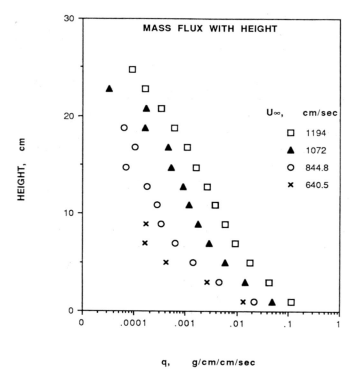

Fig. 7. The logarithm of the mass flux, q, as a function of height for different wind strengths

Nalpanis [9] had proposed that the following relationship be used to describe the exponential dependence of mass distribution with height,

$$q = \alpha e^{-[\lambda g y / u_*^2]}$$

where α and λ are constants for a single experiment. However, Nalpanis concluded based upon analysis of his data that "there does not seem to be any *a priori* way of determining the values of α and λ". The present data also was fitted to this expression, and, although there are definite tendencies of the data, i.e., λ and α both increasing with increasing u_* values, the same conclusion of Nalpanis was reached for our data. No satisfying equation could be established to account for the α and λ variation from one experiment to another. The α and λ parameters appear not to be constants but rather functions of u_*, particle density and probably functions of particle shape, distribution of particle size and surface roughness. Our values of α ranged from about 30 to 100 mg/(cm² · s), while values of λ varied from unity to about 1.5 both of which uniformly increased with increasing friction speed. The range of α and λ values somewhat agree with the Nalpanis values when the density ratio of walnut shell to sand is accounted for in the λ parameter (our λ's typically are larger by a factor of 2.5 to 3, the ratio of sand to walnut shell density).

8 Total mass flux collection

The total particle flow consists of the saltation together with the surface creep and a small remainder which may be carried in suspension. The surface creep derives its forward momentum from the bombardement of the impacting saltating particles. It is believed to be

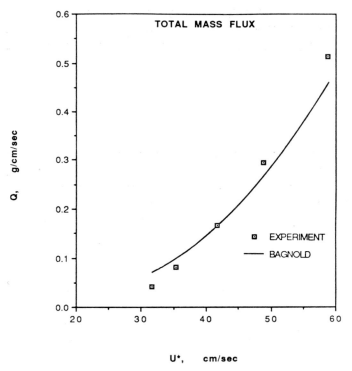

Fig. 8. Total mass flux as a function of friction speed. The solid curve is the theory of Bagnold [1]

approximately equal to a quarter of the total mass transport. The effect of particles in suspension may be safely neglected, except in the case of a very dusty particle flow (particles of 20 microns or less). Accordingly, a 120 mesh screen was used in the construction of the back and top of the total mass flux collector, which was of similar geometric shape as the Section VII collectors without the individual vertical departments, i.e., one single vertical collector which extended to the tunnel flows to the tunnel ceiling. The screens allowed air to pass through the collector and decrease the pressure build-up inside the collector. A 60° angle plate (baffle) was placed inside the collector, at half the distance from the collector leading edge. This baffle helped decrease the pressure build up inside the collector. The angle between the side plate and the perpendicular plate was varied to obtain the optimum collector design for maximum particle collection.

Three different collector angles were tested and flux measurements made and averaged. The freestream velocity was held constant at 900 cm/s during these experiments. It was found that the optimum collector had an angle of 12°, which corresponded to the same angle used in constructing the individual collectors. This value of the angle supported the results obtained for the mass flux as function of height given in Section VII.

The trapped mass was measured by emptying the collector particle mass on the Metler scale plate. The total flux, Q, consisted of the mass passing through a unit vertical width during a measured time. The measured total mass flux was compared to published theories by White [12], Kawamura [6], and Bagnold [1]. The experimental results fell in between these three theories, more closely agreeing with Bagnold's results as shown in Fig. 8. The two other theories gave higher values of the mass flux and generally are expected to be more accurate than Bagnold's equation near threshold conditions since it does not account for zero flux below particle threshold. The discrepancy between the various theories and data may be explained by the fact that it is known that not all wind-tunnel particle collection systems for total flux determination are accurate or 100% efficient, usually resulting in low mass flux measurements. The shape of the collector is very important in collecting mass fluxes.

9 Concluding remarks

The present study considered several aspects of wind-tunnel saltation including: the effect of Froude number, the determination of mean velocity profiles and friction speeds in saltation as a function of downstream position, saltating mass as a function of height and total mass flux. Most of the experimental measurements were acquired in a relatively small UC Davis boundary-layer wind tunnel. Some previously unpublished MARSWIT wind-tunnel data was also utilized. The Froude number ranged from 6 to 1000.

Some of the major findings are summarized below

— The principle of an independence Froude number (proposed by Owen and Gillette [11]), below which saltation is uninfluenced by the wind tunnel walls, is supported by the present study. However, the value of the critical Froude number may be lower than 20, the value suggested by Owen and Gillette. The present data suggest flows with Froude numbers of 10 or less obtain constant friction speed status more quickly than higher Froude number flows, although it is noted that selection of the critical Froude number is somewhat subjective and may vary among different investigators.

— Additionally, there appears to be a minimum entrance length necessary to develop equilibrium saltation, which is defined as having uniform downstream values of friction speed. Experimentally, it was observed that this distance corresponded to a tunnel length-to-height ratio of at least 5. Moreover, a minimum downstream distance of 25 boundary layer heights is suggested to insure a reasonable developed turbulent boundary layer flow, in the mean flow characteristics only, is present.

— In order to have well defined saltation parameters u' and k', as defined by Bagnold [1] it was necessary that the saltation process had reached equilibrium. This means that the Froude number must be below the critical value and x/H must be in excess of 5 for u' and k' to exist. This may explain why some previous researchers failed to achieve or observe u' and k' in their wind-tunnel saltation flows.

— The collection of total vertical mass flux was found to be a strong function of the collector geometry giving raise to the validity of these measurements without first investigating the effect of the aerodynamic efficiency of the collector. Present results were found to follow the Bagnold equation reasonably well when the optimum collector shape was used.

References

[1] Bagnold, R. A.: The physics of blown sand and desert dunes. London: Methuen (1941).

[2] Coles, D.: The law of the wake in turbulent boundary layer. J. Fluid Mech. 1, 191—226 (1956).

[3] Greeley, R., Iversen, J. D.: Wind as a geologic process on Earth, Mars, Venus, and Titan. Cambridge: Cambridge University Press, p. 100 (1985).

[4] Greeley, R., White, B. R., Pollack, J. B., Iversen, J. D., Leach, R. N.: Dust storms on Mars:

considerations and simulations. In: Desert dust: origin, characteristics, and effect on man. (Perve, T., ed.) Special on Geological Society of America, (1981).

[5] Iversen, I. D.: Saltation in the wind tunnel. Unpublished manuscript, (1988).

[6] Kawamura, R.: Study on sand movement by wind (in Japanese). Institute of Science and Technology, Tokyo University, Tokyo, Report 5, 95—112, (1951).

[7] Leach, R. N.: Private communication, (1984).

[8] Mounla, H.: Wind-tunnel measurements of mass transport and velocity profiles in saltating turbulent boundary layers. Master of Science Thesis, University of California, Davis (1989).

[9] Nalpanis, P.: Saltating and suspended particles over flat and sloping. II. Experiments and numerical simulation. Proceedings on International Workshop on the Physics of Blown. University of Aarhus, vol. I, (1985), pp. 37—66.

[10] Owen, P. R.: Saltation of uniform grains in air. J. Fluid Mech. 20, 225—242 (1964).

[11] Owen, P. R., Gillette, D.: Wind tunnel constraint on saltation. Proceedings on International Workshop on the Physics of Blown Sand. University of Aarhus, vol. 2. (1985), pp. 253—269.

[12] White, B. R.: Soil transport by winds on Mars. J. Geophys. Res. 84, B8, 4643—4651 (1979).

[13] White, B. R.: Low-Reynolds-number turbulent boundary layers. J. Fluid Engng. 103, 624—630 (1981).

[14] White, B. R.: Two-phase measurements of saltating turbulent boundary layer flow. Int. J. Multiphase Flow 5, 459—473 (1982).

[15] White, B. R.: Particle dynamics in two-phase flows. Chapter 8 of Encyclopedia of fluid dynamics. Houston, Texas: Gulf Publishing Co., (1986) pp. 239—282.

[16] White, Frank M.: Fluid mechanics, 2nd ed. McGraw-Hill (1986), p. 398.

[17] Zingg, A. W.: Wind tunnel studies of the movement of sedimentary material. Proceedings of 5th Hydraulic Conference, Bulletin 24, 111—135 (1953).

Authors' address: B. R. White and H. Mounla, Department of Mechanical, Aeronautical and Materials Engineering, University of California, Davis, CA 95616, U.S.A.

Acta Mechanica (1991) [Suppl] 1: 159—166

Grain shape effects on aeolian sediment transport

M. A. Rice, Aberdeen, United Kingdom

Summary. The influence of shape as shown by photographic studies of collision, aerodynamic entrainment and grain trajectories is discussed as a factor in determining sediment transport rate. Collision generates more bed activity as sphericity increases, whereas grains with a low sphericity are more readily entrained at slow windspeeds by direct action of the wind. As sphericity decreases grain trajectories become longer and flatter.

Introduction

In the literature of wind-blown sand there are many measurements of mass transport rate for sands of apparently similar size and density. Some of these differ significantly. Most prediction formulae, such as that shown in equation 1 (Banglod [1]), incorporate grain size and shear velocity.

Mass transport rate,

$$q_{,} = C \left(\frac{d}{D} \right)^{1/2} \frac{\varrho}{g} U_{*}{}^{3} \tag{1}$$

C is a coefficient, d is the median sand diameter, D is the diameter of a standard 0.25 mm sand, ϱ is the air density and U_{*} is the shear velocity. The coefficient, C, takes into account the degree of sorting present in the sand population.

The discrepancies found in measurements of transport rate appear rather large to be accounted for by different sorting characteristics alone and may be influenced by a variety of other factors (either singly or in combination). It is probable that some differences arise from experimental errors, such as those derived from poorly designed traps, inaccurate measurement of the wind velocity profile, and grain size determination. Other possible factors include grain density, progressive sorting, shape, packing, aggregation, moisture, surface crusts and relative humidity. Of these, grain shape and density are properties of the particles and therefore may be considered the most important intrinsic parameters after "size". The nature of packing and progressive sorting depends on these particle properties. The remaining factors are capable of exerting considerable influence on the transport rate, but they arise from external conditions superimposed upon the characteristics of particular populations.

The effects of grain density and grain shape have not been studied extensively in air. Sedimentary particles commonly have densities which vary little from that of quartz, so this parameter can usually be neglected, unless heavy minerals are significant in the grain population. Grain shape has usually been ignored in wind transport, partly because it was not considered to be important and partly because it is difficult to measure accurately and quickly.

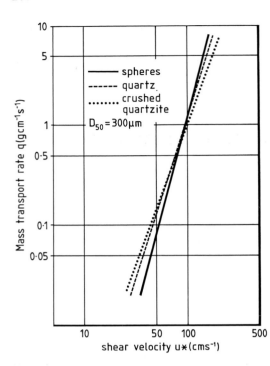

Fig. 1. Transport rates found by Williams (1964) for populations with different shapes

Previous studies concerned with grain shape effects on aeolian transport rate produced conflicting results. Some workers, e. g. [2], [3], suggested that the wind carries more spherical[1] grains faster and further than angular grains. Others, including Mattox [5] and Winklemolen [6] postulated that the wind should select grains of lower sphericity during saltation and suspension.

Methods of shape measurement differed considerably in the studies cited above, as did the location and method of sampling. In none of the experiments was the wind condition monitored or discussed.

A clearer picture emerges when shape influence on transport rate is examined in the controlled environment of a wind tunnel. Williams [7] investigated the movement of separate grain populations of glass spheres, natural quartz and crushed quartzite. He found that at wind speeds not greatly above threshold, transport rate increased with a decrease in particle sphericity, while at wind speeds well above threshold the opposite effect occurred (Fig. 1). Willetts, Rice and Swaine [8], working with a natural quartz dune sand and a platy shell sand reached similar conclusions.

Sand sized grain populations move largely by saltation and reptation/creep. Saltation, being the dominant mode of transport, has the greatest influence on the manner and rate of movement of grain populations.

This paper investigates the influence that shape has on saltation, using high-speed and video photography of individual particles in a wind tunnel. A description of the tunnel can be found in Willetts [9]. Observations were made of the behaviour of two sands, one quartz the other shelly. They have been described in detail by Willetts and Rice [10], [11]. Briefly, the two populations have similar sieve-size distributions and mineral density, but different

---

[1] They actually measured "roundness" a second level shape property but other workers, including Mattox, consider that the methods they used were really estimating "sphericity", a measure of overall particle form, and not roundness

characteristic grain shape (as measured by the Sneed and Folk method [12]). The quartz sand and the shelly sand have mean sphericities of 0.73 and 0.59 respectively.

Earlier work used high speed photography to look at individual intersaltation collisions over horizontal and sloping beds [11], [13], [14]. This study concentrates on shape effects on aerodynamic entrainment and on grain trajectories in addition to the shape effects observed for collision.

Intersaltation collisions on horizontal and sloping beds

High-speed photography of individual collisions was analysed to produce information on incoming velocities and angles (V_1, α_1), ricochet velocities and angles (V_3, α_3) and velocities and angles (V_n, α_n) for all associated grain ejections. Comparison of the horizontal and sloping bed data for quartz and shelly sand led to the following deductions.

(a) Approach and rebound angles decrease with a decrease in sphericity.
(b) The velocity ratio, V_3/V_1, decreases with a decrease in sphericity and with increasing bed slope.
(c) There is no systematic change in either mean ejection velocities or ejection angles relative to the bed with particle shape.
(d) Forward momentum loss per collision (i.e. momentum lost parallel to the bed slope to creeping grains) is greater for the quartz sand over a horizontal bed. Momentum loss increases for both sands with increase in bed slope, by whether the rate of increase changes with particle shape is not clear.
(e) Change in vertical momentum increases with an increase in particle sphericity and with an increase in bed slope.
(f) The number of ejections per collision and the percentage of collisions causing dislodgement, increase with increasing sphericity. This activity increases with bed slope for the quartz sand. However, for the shelly sand, an increase in bed angle does not affect the percentage of dislodging grains and appears to inhibit bed activity in terms of numbers of ejecta.

The collision experiments display a marked difference between the two shapes studied. Platy grains promote far less bed activity than the more spherical quartz grains over both flat and rippled beds. In almost all other respects the quartz sand also contributes to and is affected by the collision much more than the shelly sand. Thus the importance of collision increases with increase in sphericity.

High speed films of aerodynamic entrainment

At windspeeds above threshold, aerodynamic entrainment was observed near the leading edge of a deposit, where grain detachment from the bed cannot be the result of upwind grains saltating into the area. The high-speed camera was positioned at 90° to and on a level with the length of a 5×30 cm sand strip set longitudinally on the centre line of the tunnel floor. Fixed surface roughness of the same particle size existed for 3 m upwind.

Visual observations indicated that the first movement of particles appeared to be small hops. Further downwind the hops increased in height and the number of grains in motion increased rapidly. The downwind portion of the strip was progressively eroded, the rate of removal increasing with increasing windspeed. However, the upwind edge of the strip remained more or less intact over most of the range of windspeeds tested. The optimum field of view for good grain definition was 10 cm. It was therefore impossible to follow a grain from incipient movement near the leading edge of the strip, through to hopping and to saltation capable of ejecting other grains. It appears that this length is approximately 20—30 cm.

The upwind 10 cm portion of the strip was examined. Three runs at different windspeeds were photographed at a filmspeed of approximately 1500 frames per second. The lowest windspeed corresponded to that which initiated patches of grains on the upwind part of the strip into motion. Two slightly higher windspeeds followed.

Data analysis of the films was achieved by digitization of individual grain paths from initial movement until the particle left the field of view. Grain paths were variable. Some rolled for several centimeters before performing a hop. Others left the bed after a short rolling movement. Two hops over the length of the bed were common. A variable proportion (5—28%) of grains left the bed instantaneously with no previous rolling movement. No grain oscillation was observed prior to take-off. The hopping grains were seen to tumble in the air, the spinning motion continuing in the fashion of the rolling grains. This forward spinning was also evident in the grains which lifted directly off the bed without rolling.

Direct lift-off of individual grains occurred more frequently for the shelly sand. In some films multiple spontaneous lift-off occurred near the leading edge of the deposit, whereby

Table 1. Aerodynamic entrainment. Mean velocities, angles and lengths

	Q	S	Q	S	Q	S
U_* (cm s^{-1})	44	44	51	51	59	59
No. of grains	23	72	82	76	57	75
% direct take-off	8.7	26.8	27.8	21.1	5.3	8.0
Lift-off x co-ordinate	1.0	1.3	1.0	0.8	0.5	1.2
1st jump x co-ordinate (— L. O.)	2.7	2.9	2.9	2.7	2.4	2.5
Rolling velocity prior to 1st jump (cm s^{-1})	44.2	56.5	48.8	66.1	48.1	60.4
Lift-off velocity (cm s^{-1})	49.5	54.9	42.8	56.8	44.7	58.5
1st jump velocity (— L. O.) (cm s^{-1})	44.2	62.1	46.9	70.8	51.0	68.1
Lift off angle (°)	18.4	26.3	24.4	12.9	18.5	12.9
1st jump angle (— L. O.) (°)	25.3	14.6	21.3	13.8	17.7	13.6
Rolling length before 1st jump (cm)	1.0	1.2	1.1	1.1	0.7	1.0
First jump length (cm)	1.4	2.0	1.9	2.3	1.8	1.6

Q quartz sand; S shelly sand
— $L. O.$ minus direct lift-off grains; x *co-ordinate* distance from leading edge of strip in cm

up to 20 neighbouring particles suddenly left the bed. This phenomenon will be referred to as a "flurry". Some of these grains disappeared from the field of view without striking the bed again. Others performed a hop or two. Again no grain oscillation prior to lift-off was observed. The "flurries", which occur in both sands, obscure comparisons beetween single grains lifting off, which are more common for the shelly sand. "Flurry" frequency is impossible to assess because of the short duration of the films (3.2 s). They occur on only half of the films.

At each windspeed the shelly sand was more mobile than the quartz sand, particularly at the lowest speed. Most of the rolling began at or close to the leading edge.

The films produced information on the progress of grains moving downwind in relation to the leading edge of the deposit in terms of distance travelled, speeds, angles and hop lengths. Comparisons between the quartz and shelly sands are given as mean values in Table 1.

Differences between the two sands occur in rolling and jump velocities and in jump angles for similar windspeeds. Rolling and first jump velocities are always higher for the shelly sand as are the initial speeds of grains which lift-off directly from the bed. Take-off angles for grains which previously rolled are normally shallower for the platy sand, but no shape influence was observed for angles of grains which lifted-off directly. Mean rolling lengths were greater for the shelly sand.

Usually the trajectory length for the first hop is longer for the shelly sand. Lengths gradually increase with increase in windspeed. As would be expected, there is a general increase in hop length for the second and third jumps (not shown in Table 1).

Trajectories

A video film of saltation showed small trajectories (< 16 cm) and traces of longer trajectories. At similar wind speeds the shelly sand travels a longer, shallower path. As has already been noted it produces much less bed activity, perhaps partly because longer trajectories imply fewer collisions.

The interaction of particle shape with the wind, particularly in relation to the nature of the trajectory and to grain velocities relative to the wind, will be influenced by drag forces. The equation for the total instantaneous drag on a single grain

$$F_D = C_D \frac{1}{2} \varrho_a U^2 A \tag{2}$$

depends upon the drag coefficient, C_D, the area of the cross-section of the particle presented to the wind, A, and the relative velocity between the grain and the air, U. As the shape deviates from a sphere, the surface area increases (for similar mass) and the drag coefficient

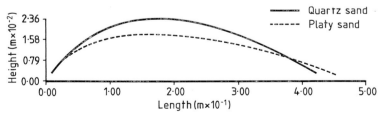

Fig. 2. Computer simulations of grain trajectories. Projected areas based on a diameter of 300 μm (quartz) and 500 μm (platy sand)

increases [15]. At the start of a trajectory, spinning commonly occurs about a horizontal axis perpendicular to the direction of flow. For the quartz sand, rotation does not substantially alter the area presented to the flow. Rotation of the platy sand continually changes the area, A, in equation 2 and this may cause variations in the total drag force.

Figure 2 shows trajectories simulated by McEwan and Willetts' saltation model (see this volume [16]). Compared with a near-spherical quartz grain, an increase in the average projected area of the platy grain throughout the trajectory shows a substantial change in the shape of the trajectory (Fig. 2) to a longer and flatter path. Lower mean ricochet angles, always seen with the platy sand, will enhance this effect. The longer, lower trajectories observed on the video films for the shelly sand are therefore consistent with an increased drag force due to the shapes of the grains.

Discussion

It is evident from these studies that shape does influence the movement of sand sized particles in an aeolian environment.

Collision has been shown to be far more significant for the more spherical sand. These grains rebound more vigorously than angular grains, impart more momentum to creeping/reptating grains and generate more dislodgements involving more ejecta. Once collision becomes the dominant mechanism for releasing bed grains into the airflow, then spherical grains would be expected to interact more efficiently with the bed. At moderate to high windspeeds when collision predominates, this would produce a higher transport rate. Thus the difference in transport rate observed at these windspeeds by Williams [7] and Willetts et al. [8] can at least partly be attributed to collision.

As sediments start to move, aerodynamic processes are more important than collision to the initial motion and entrainment of grains. Evidence, both from leading edge films and from visual observation of initial grain movement in the wind tunnel, suggests that angular grains are mobilized by direct wind action more readily than spherical grains. Particles may begin to move when the lift force (or a combination of lift and drag) exceeds the opposing force due to the weight of the grains. It has also been suggested by Williams et al. [17] that the mechanism for launching surface particles into the air is similar to that reported for water [18], [19]. They consider that the "bursting" process of a turbulent boundary layer is a likely vehicle for entrainment of surface grains. Surface roughness is an important factor in the magnitude and frequency of burst sequences [20]. Although the difference in surface roughness between the two sands is small in terms of sieve size, the marked difference in grain shape means that the surface packing will vary. Bulk densities indicate that the shelly sand is less well packed and therefore contains a larger volume of pore space than an equivalent layer of quartz sand. Grass [20] suggested that low momentum fluid trapped between roughness elements acts as a reservoir for the bursting process. In addition to the potentially larger reservoir, the overlap of the probability distribution of the local instantaneous shear stresses with the probability distribution of grain susceptibility, [19], is likely to be larger due to variations in bed texture and a consequent increase in the vertical velocity fluctuations. Therefore an angular grain may be more easily entrained by direct lift-off by the wind than a spherical grain of equivalent size.

The video films show that once saltation is established the grain trajectories are still influenced by shape. The platy grains follow a longer and flatter path than the more spherical grains. At a similar windspeed they are still travelling faster as evidenced by the longer traces on a single video frame. If the dislodgement rate time constant (K) and the number of exposed grains per unit area (No) is known, then the transport rate can be assessed for different distributions of path length. For similar values of KNo, the transport rate will be higher for longer path lengths. Thus an increased concentration of platy grains at low windspeeds may be due to longer trajectory lengths in addition to their enhanced entrainment. As collision starts to build up the probability of dislodgement of exposed grains can be expected to change. Once a grain has started to jump, there is a high probability that it will continue to do so and eject other grains at collision.

Willetts [9], who used the two sands discussed in this study, reported a discontinuity in the transport rate for the quartz sand when measured against point velocity at several heights. Similar behaviour was also observed for a more dense material of comparable shape, but was absent in the results of the platy sand. Willetts discounted any link with ripple formation, for ripples exist comformably at windflows above and below the observed discontinuity. He speculated that the phenomenon could be associated with a transition between different sediment transport mechanisms. Only one mechanism, wind dislodgement, was thought to be predominant over the range of windspeeds studied for the shelly sand.

It is possible that the low collision efficiency of shelly sand never allows the impact process to become the dominant mechanism for dislodgement. However, the transport rate can only be maintained if sufficient dislodgements are generated to replace terminating saltation sequences. With the platy sand this may be achieved by aerodynamic entrainment in balance with newly ejected grains and terminating saltating grains. Alternatively, aerodynamic entrainment may not necessarily be significant over the whole range of windspeeds. The video films leave an impression that the longevity of the platy grains is greater than the quartz ones, i.e. they survive for a greater number of hops. It could be speculated that once airborne, the platy sand continues in motion for longer than the quartz sand, which suffers greater momentum loss upon collision.

Willetts also reported that for similar tunnel settings the quartz sand retards the wind velocity profiles close to the bed far more than the shelly sand does. The more spherical sand extracts momentum from the wind more successfully than the platy sand. The smaller effect that the platy grains have on near bed velocities means that the capacity for aerodynamic entrainment, although diminished, is likely to remain.

Conclusions

Photographic techniques have shown that grain shape is influential in the movement of sediment by wind in the following ways.

(1) The nature of collision is influenced by particle shape. With a decrease in sphericity rebound is less vigorous, less momentum is imparted to creeping/reptating grains and fewer dislodgements involving fewer ejecta are generated.

(2) With a decrease in sphericity, impact, ricochet and ejection angles are shallower.

(3) With decrease in sphericity saltation trajectories become longer and flatter.

(4) There is evidence that as the shape of a particle deviates from a sphere its probability of being entrained aerodynamically increases.

(5) The near-bed wind is influenced less by the less spherical grains because the impacts remove less forward momentum. Therefore the wind retains greater entrainment capability in the case of the less spherical material.

These observations are consistent with the results of Williams and Willetts, Rice and Swaine, and provide a possible explanation for mass transport rate differences at low and high windspeeds. They are particularly pertinent in areas of coastal dunes, which normally consist predominantly of either quartz or shell fragments.

Acknowledgements

This study was supported by the Natural Environment Research Council. Grateful thanks are extended to I. McEwan and B. Willetts for their helpful comments.

References

[1] Bagnold, R. A.: The physics of blown sand and desert dunes. London: Methuen, pp. 265 (1941).

[2] MacCarthy, G. R., Huddle, J. W.: Shape-sorting of sand grains by wind action. Am. J. Sci. **35**, 64—73 (1938).

[3] Shepard, F. P., Young, R.: Distinguishing between beach and dune sands. J. Sedimentary Petrol. **31**, 196—214 (1961).

[4] Barrett, P. J.: The shape of rock particles: a critical review. Sedimentology **27**, 291—303 (1980).

[5] Mattox, R. B.: Eolian shape-sorting. J. Sedimentary Petrol. **25**, 111—114 (1955).

[6] Winklemolen, A. M.: Experimental rollability and natural shape sorting of sand. Ph. D. Thesis, Groningen, Netherlands (1969).

[7] Williams, G.: Some aspects of the eolian saltation load. Sedimentology **3**, 257—287 (1964).

[8] Willetts, B. B., Rice, M. A., Swaine, S. E.: Shape effects in aeolian grain transport. Sedimentology **29**, 409—417 (1982).

[9] Willetts, B. B.: Transport by wind of granular materials of different grain shapes and densities. Sedimentology **30**, 669—679 (1983).

[10] Willetts, B. B., Rice, M. A.: Practical representation of characteristic grain shape of sand: a comparison of methods. Sedimentology **30**, 557—565 (1983).

[11] Willetts, B. B., Rice, M. A.: Collisions in aeolian saltation. Acta Mech. **63**, 255—265 (1986).

[12] Sneed, E. D., Folk, R. L.: Pebbles in the Lower Colorado River, Texas. A study in particle morphogenesis. J. Geol. **66**, 114—150 (1958).

[13] Willetts, B. B., Rice, M. A.: Insaltation collisions. Proc. Int. Workshop on the Physics of Blown Sand. Aarhus, pp. 83—100 (1985).

[14] Willetts, B. B., Rice, M. A.: Collisions of quartz grains with a sand bed: the influence of incident angle. Earth Surface Processes and Landforms **14**, 719—730 (1989).

[15] Komar, P. D., Reimers, C. E.: Grain shape effects on settling rates. J. Geol. **86**, 193—209 (1978).

[16] McEwan, I. K., Willetts, B. B.: Numerical model of the saltation cloud (this volume).

[17] Williams, J. J., Butterfield, G. R., Clark, D. G.: Aerodynamic entrainment threshold: effects of boundary-layer flow conditions (in press).

[18] Kline, S. J., Reynolds, W. C., Schraub, F. A., Runstadler, P. W.: The structure of turbulent boundary layers. J. Fluid Mech. **30**, 741—773 (1967).

[19] Grass, A. J.: Initial instability of fine bed sand. Proc. J. Hydraul. Div. A.S.C.E. **96 (HY3)**, 619—632 (1970).

[20] Grass, A. J.: Structural features of turbulent flow over smooth and rough boundaries. J. Fluid. Mech. **50**, 233—255 (1971).

Author's address: M. A. Rice, Department of Engineering, University of Aberdeen, Kings College, Aberdeen AB9 2UE, United Kingdom.

Acta Mechanica (1991) [Suppl] 1: 167—181

Two- and three dimensional evolution of granular avalanche flow – theory and experiments revisited

K. Hutter, Darmstadt, Federal Republic of Germany

Dedicated to Prof. Y.-H. Pao on the occasion of his sixtieth birthday

Summary. Rockfalls, sturzstroms, landslides and snow and ice avalanches have been conjectured to be describable by a plastic continuum of the Mohr-Coulomb type with a constant internal angle of friction. Bed resistance is modeled through a sliding law by additively composing the basal traction of a Mohr-Coulomb stress with bed friction angle δ and a viscous drag that is proportional to the first or second power of the sliding velocity. The theoretical model takes advantage of depth averaging and presents field equations for the height and surface-parallel volume fluxes.

Results are reviewed that have been obtained with this theory in chute flows along straight and curved beds and unconfined motions of a finite mass of a cohesionless granular material down plane and curved beds. The results are compared with experimental findings from extensive laboratory tests, and inferences are drawn that derive from such comparisons.

1 Theoretical concept

The classical models that describe the motion of avalanching rock, snow and ice masses down mountain sides are based on the simple hydraulic or particle concepts of Voellmy [1], Salm [2], Perla et al. [3] and others. By proper adjustments of the 'turbulent friction coefficient these models were employed for the dynamics of both dense flow and air borne powder snow avalanches, but they are now gradually being extended or superseded by two separate classes of theories that describe the respective physical situations in more detailed and complete ways.

Attention here is confined to the gravity driven rock, ice and dense flow snow avalanches which are not driven by the turbulence of the air, but are rather governed by the interactions between the particles that cause the granular mass to behave in a fluid-like fashion. Several hypotheses have been proposed to explain the mechanisms for this fluidization, see e.g. the references in [4], [5]. Kinetic theory and molecular dynamics models — the most convincing among these — have been developed to describe the constitutive behaviour of granular materials, see again [4], [5] for references. However, as demonstrated by Hutter et al. [6], [7] these models are cumbersome to handle even when applied to such a simple physical situation as plane shear flow. Fortunately, a much simpler 'hydraulic' model with a Coulomb-type basal friction law is likely to be sufficient to predict the main features of these flows, and the discussion here will describe the progress that has been achieved on this front in recent years.

The subject of this study is the two- or three-dimensional motion of a pile of cohesionless granular material as it starts from rest and flows down a rough bed. The bed is assumed to have a steep slope at the initial position of the pile and may be curved in a downhill direction, so as to approach a horizontal flat. In the other direction curvature effects will be ignored, so that under unconfined three-dimensional flow maximum sidewise spreading is achieved. Two-dimensional plane flow situations are obtained when the flow mass is of infinite extent perpendicular to the main flow direction. We have approximately achieved it by confining the moving granular mass between two vertical walls. Transverse shear could not completely be avoided, but the flow dynamics was corrected for it by incorporating wall friction from the sides.

Even though the avalanche is made up of discrete granules it shall in all ensuing problems be treated as a *continuum*. Thus, depths and lengths must be large compared to the dimensions of a typical particle. The continuum is assumed to be *incompressible* and satisfies the constitutive postulates of a *Mohr-Coulomb plastic material* with constant angle ϕ of internal friction. Likewise, a Mohr-Coulomb yield criterion will be employed at the basal surface, which will model the *sliding* processes at the bed and relate the shear traction at the bottom to the normal stress by a constant *bed friction angle* $\delta < \phi$. These assumptions are idealized ones, but under 'usual' situations they have been motivated and justified to be adequate approximations in [6], [8] and [9].

It is not entirely clear what practitioners mean by the term 'usual'. In short, the conjecture seems to be that to the Mohr-Coulomb basal drag a *viscous component* must be added to fit the simple models with observations. This additional drag will be referred here to as *Voellmy drag*. Its addition has been studied in [10]; it, of course, introduces a further phenomenological coefficient into the theory. We have found that for plane flow this *Voellmy* drag need not be introduced to obtain a better agreement between theory and experiment: only when avalanche flows are three-dimensional does it seem needed for a better match between theory and experiments. Other, more complex, basal resistance hypotheses have not been studied so far.

The spatially two- and three-dimensional motion of this idealized plastic flow model is further conceptionally simplified by restricting consideration to granular geometries that are long, shallow and basically slowly varying. Furthermore, the basal sliding velocity is large and the internal shear deformation is generally small. This justified the introduction of *depth averaged equations*. The governing field quantities are then the height of the granular pile and the depth averaged velocity components in one or two directions parallel to the bed (one longitudinal in the direction of the main motion, the other lateral or sidewise, when unconfined motion is possible). Furthermore, the shallowness assumption permits development of the equations in terms of a small aspect ratio parameter ε which equals a typical depth divided by a typical lateral or longitudinal dimension. The present model equations incorporate terms involving ε up to linear order. And, finally, the slow variation also restricts curvatures of the bed to be small in the sense that the bed radius of curvature be at most as large as the granular pile length.

The evolution of the motion of the granular pile of this model depends upon *two*, possibly *three physical parameters*, the internal angle of friction, ϕ and the bed-friction angle, δ and possibly a viscous drag coefficient. The remaining quantities are then only geometrical and involve the form and size of the initial shape (determining for instance ε) and the geometry of the basal surface.

2 Chute flow models

Here attention is confined to chute flows along curved beds between two parallel sidewalls.

2.1 Governing equations

Let (x, y) be Cartesian and (ξ, η) orthogonal curvilinear coordinates in a vertical plane as shown in Fig. 1a. The ξ-line corresponding to $\eta = 0$ follows the basal profile, the η-lines are straight rays perpendicular to the basal profile. To span the avalanche domain uniquely by this coordinate system the depth of the sliding mass cannot exceed $r(\xi)$, the radius of curvature of the basal profile, an assumption that will always be made in the sequel. The avalanche depth and the depth averaged velocity component in the ξ-direction will be denoted by $h(\xi, t)$ and $u(\xi, t)$, respectively. The geometry of the known sliding curve may be expressed in terms of the Cartesian coordinates x, y; the relationship allows the bed inclination angle ζ and the local curvature $\varkappa$ of the basal surface to be expressed as functions of ξ: $\zeta = \zeta(\xi)$, $\varkappa = \varkappa(\xi)$. In what follows, lengths, time and velocity are to be interpreted as dimensionless, scaled in accordance with the physical problem at hand, see Table 1.

Savage and Hutter [4] have shown that in these dimensionless variables the equations of depth averaged mass and momentum balances take the form

$$(1 - \varepsilon h \lambda \varkappa) \frac{\partial h}{\partial t} + \frac{\partial (hu)}{\partial \xi} = 0, \tag{2.1}$$

$$\left(1 - \frac{\varepsilon \lambda \varkappa h}{2}\right) \frac{\partial u}{\partial t} + \left(1 + \frac{\varepsilon \lambda \varkappa h}{2}\right) u \frac{\partial u}{\partial \xi} = \left(1 - \frac{\varepsilon \lambda \varkappa h}{2}\right) \sin \zeta - \frac{\varepsilon h}{2} \frac{\partial}{\partial \xi} \left(k_{\text{actpass}}(\cos \zeta + \lambda \varkappa u^2)\right)$$

$$- (\cos \zeta + \lambda \varkappa u^2) \left[\left(1 + \frac{\varepsilon \lambda \varkappa h}{2}\right) \operatorname{sgn}(u) \tan \delta + \varepsilon k_{\text{actpass}} \frac{\partial h}{\partial \xi}\right]$$

in which the earth pressure coefficient is given by

$$\left. \begin{array}{c} k_{\text{act}} \\ k_{\text{pass}} \end{array} \right\} = 2 \left[1 \mp \sqrt{1 - (1 + \tan^2 \delta) \cos^2 \phi}\right] \Big/ \cos^2 \phi - 1, \quad \text{for} \quad \partial u / \partial x = \begin{array}{c} > 0 \\ < 0. \end{array} \tag{2.2}$$

Here, $\varepsilon = H/L$ is the aspect ratio and λ is a typical scale of curvature of the bed. Equations are valid for $\varepsilon \ll 1$ to linear order in this parameter and it was assumed that $\varepsilon < O(\tan \delta) < 1$ and $\varepsilon \leqq O(\lambda) \leqq 1$.

Since realistic values of δ are between $10°$ and $30°$, the first of these does not pose any restrictions beyond the shallowness assumption, the second supposes moderately weak curvature in the sense that the radius of curvature along the bed should everywhere exceed the length scale L. The equations incorporate the transverse momentum balance.

Equations (2.1) comprise a system of two partial differential equations for the depth profile $h(\xi, t)$ and the depth averaged velocity $u(\xi, t)$ for the avalanche, however they have not been solved in this generality, but where further simplified by making different choices of ordering for the parameters $\tan \delta$ and λ and then neglecting terms of order higher than ε:

(i) $\tan \delta = O(\varepsilon^{1/2})$ and $\lambda = O(\varepsilon^{1/2})$. These assumptions are often justified. With them, the equations (2.1) become

$$\frac{\partial h}{\partial t} + \frac{\partial (hu)}{\partial \xi} = 0,$$

$$\frac{\partial u}{\partial t} = \frac{\partial u}{\partial t} + u \frac{\partial u}{\partial \xi} = \sin \zeta - \tan \delta \operatorname{sgn}(u) (\cos \zeta + \lambda \varkappa u^2) - \varepsilon k_{\text{actpass}} \cos \zeta \frac{\partial h}{\partial \xi}. \tag{2.3}$$

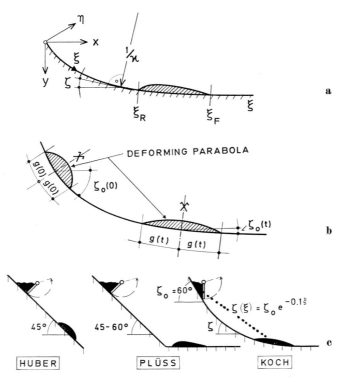

Fig. 1. a Definition sketch of coordinate system and geometry of a finite mass of granular material moving down a rough bed. (x, y) are Cartesian coordinates, (ξ, η) are curvilinear. $\zeta(\xi)$ denotes the inclination angle of the bed an $\varkappa(x)$ the curvature. ξ_r and ξ_f are the rear and front end of the avalanche. **b** Parabolic shape of the granular avalanche on a weakly curved bed. Shown are the initial position with the semi-width $g(0)$ and the midslope angle $\zeta_0(0)$, and a general position with the corresponding quantities indicated. The area of the parabolas is conserved. **c** Sketches of the arrangements of the chute experiments of Huber [13], Plüss [14] and Koch [15]. In the latter case the base is described by $\zeta = \zeta_0 \, e^{a\zeta}$, with $\zeta_0 = 60°$, $a = 0{,}1$. The dotted line indicates the 'Pauschalgefälle' with inclination angle δ

Table 1. Scales for the nondimensional variables

— Lengths parallel to the basal surface, ξ:	L, L_x, L_y
— Depth function, h:	H
— Curvature of the bed, α:	$\dfrac{\lambda}{L}$
— Velocities parallel to the bed, u, v:	$\sqrt{gL}$
— Time, t:	$\sqrt{L_x/g}$

(ii) $\tan \delta = O(\varepsilon)$ and $\lambda = O(\varepsilon^\alpha)$, $\alpha > 0$. Here we have

$$\frac{\partial h}{\partial t} + \frac{\partial(hu)}{\partial \xi} = 0,$$

$$\frac{\partial u}{\partial t} + u \frac{\partial u}{\partial \xi} = \sin \zeta - \tan \delta \cos \zeta \, \mathrm{sgn}\,(u) - \varepsilon k_{\mathrm{actpass}} \cos \zeta \frac{\partial h}{\partial \xi}.$$

(2.4)

The interpretation of these equations is as follows: In their most general form (2.1) the local and convective time derivatives contain prefactors which are curvature dependent. They contain the product $\varepsilon\lambda$ of the scales ε and λ and are the manifestation of the nonconstant metric of the curvilinear coordinates. Whenever $\varepsilon\lambda < O(\varepsilon)$ these terms can be dropped. In the momentum equations the right hand sides comprise of a series of terms each of which is interpretable as a force as follows:

- $\left(1 - \dfrac{\varepsilon\lambda\varkappa h}{2}\right)\sin\zeta$: gravity force component in the ξ-direction and driving the motion.

- $-\dfrac{\varepsilon h}{2}\dfrac{\partial}{\partial\xi}\left[\varkappa_{\mathrm{actpass}}\left(\cos\zeta + \lambda\varkappa u^2\right)\right]$: Longitudinal gradient of the pressure forces in the ξ-direction. (Notice that $(\cos\zeta + \lambda\varkappa^2)$ is the overburden pressure enlarged by the centrifugal force, so that the bracketed term defines the pressure $p_{\xi\xi}$. The factor $1/2$ is due to transverse averaging).

  It is seen that longitudinal pressure variations give rise to a resistive force. It is, however, weak and drops out in the cases (i) and (ii), see [4].

- $-(\cos\zeta + \lambda\varkappa u^2)\left[\left(1 + \dfrac{\varepsilon\lambda\varkappa h}{2}\right)\mathrm{sgn}\,u\tan\delta\right]$: Basal resistive drag directly expressing the Coulomb law of basal friction.

- $-(\cos\zeta + \lambda\varkappa u^2)\,k_{\mathrm{actpass}}\,\varepsilon\dfrac{\partial h}{\partial\xi}$: Longitudinal pressure force due to nonvanishing depth gradients. This force has an $O(1)$-contribution and is responsible for the *dispersive mechanism*; indeed, in the back part of a pile one usually has $\partial h/\partial\xi > 0$ and the force is a drag, in the frontal part one has $\partial h/\partial\xi < 0$, corresponding to a pull.

Systems (2.3) and (2.4) have been transformed to finite difference form for numerical integration, [4], [8]. Solutions obtained with these will be shown below. The reader should be cautioned that deduction of a numerically stable finite difference scheme for the evolving free boundary value problem has not been easy; solutions of such problems are generally fraught with difficulties.

2.2 Similarity solutions

Imagine the situation that a finite mass of granular material has initially *parabolic shape* (Fig. 1b) and then be released to evolve. Does this mass, upon release, perhaps, preserve its shape, thus remain a parabola and only change its aspect ratio? It has been shown, [8], that for vanishing curvature $\varkappa = 0$ (for which case the three systems above reduce to only one) such *similarity solutions* do indeed exist. The forms of the preserving geometries are indeed parabolas and the evolution of their length is described by a system of nonlinear ordinary differential equations in time.

When the basal surfaces are curved, then none of the above systems of equations admits similarity solutions exactly, but for weak curvature — system (2.4) — the $\sin\zeta$ and $\cos\zeta$ arising in (2.4) may be expanded about the moving avalanche midpoint and only terms linear in the curvature parameter λ be retained, a condition equivalent to $\lambda = O(\varepsilon)$. Such solutions were constructed for a constant bed friction angle by Savage and Nohguchi, [11], and for variable bed friction angle by Nohguchi et al. [12][1].

---

[1] These involve a further mild approximation

The main conclusions that are deduceable from the computations in these works listed below, however, are merely computational results which demonstrate useful qualitative physical behaviour but do not indicate whether similarity solutions could in any way be suitable for prognostic purposes.

2.3 Qualitative inferences from similarity solutions of plane chute flows

Here I summarize the findings given in [9], [10], [11] and [12]. Focus is thereby on qualitative features and their dependencies upon the governing parameters, ϕ, δ, and geometry. To understand the inferences, I remark that the basal shear traction τ will have a Mohr-Coulomb component τ_C and a viscous component τ_V, $\tau = \tau_C + \tau_V$, such that

$$\tau_C = \tan \delta p_\perp \qquad \text{and} \qquad \tau_V = Cu^2 \qquad\qquad (2.5)$$

in which $p_\perp$ is the pressure normal to the bed, u is the sliding velocity and C the viscous drag coefficient. Interesting in connection with the findings is that snow avalanches which have very small aspect ratios (ε app. 5×10^{-2} or even smaller) and move down a fairly plane mountain side, do often move with little dispersion and fairly constant shape, resembling 'rigid body' behaviour. They have, however, never parabolic shape. By 'rigid body' motion a parabolic shape is meant of which the size is maintained in the course of the motion.

1) The existence of a rigid body motion of the moving pile requires either a variable bed friction angle ($\tan \delta$ is larger in the front than in the back) or a bed with gradually varying inclination angle or both. An avalanche on a plane bed with constant bed friction angle and without a viscous sliding term τ_V will extend and accelerate for ever, [8], [10].

2) Along a plane bed the centre of mass can only reach a final steady speed when a viscous type drag is accounted for *and* the bed friction angle decreases from the front to the rear margin. Under such circumstances steady motion of the centre of mass and rigid body motion of the pile occur simultaneously, [10].

3) Along curved beds the existence of rigid body motions does not require a viscous drag or a variable bed friction angle. It is a delicate balance between (i) the curvature dependent driving stresses and (ii) longitudinal pressure gradients which on concave beds have competing influences. Rigid body motions may arise in both accelerating and decelerating phases of the centre of mass motion, [10], [11].

4) The smaller the initial aspect ratio ε is, the closer will be the motion of the parabolic pile to being rigid.

5) On a curved bed, when only a Mohr-Coulomb bed friction is active, $\tau = \tau_C$, then the centre of the avalanche at rest can be found by a line drawn from the centre of the parabolic pile in the initial position having slope $\tan \delta$ and intersecting it with the curve of the bed, Fig. 1c, [9], [15]. This intersection determines the centre of the parabolic pile at rest. This thumb rule of the 'Pauschalgefälle' is well known to practitioners who infer it from the rigid-body model, [18].

2.4 Comparison of finite difference solutions with experiments

Experiments on the flow of a finite mass of granular materials down inclined and curved chutes were performed by Huber [13], Plüss [14] and Koch [15] and involved the configurations sketched in Fig. 1c. The initial theoretical formulation of Savage and Hutter [8]

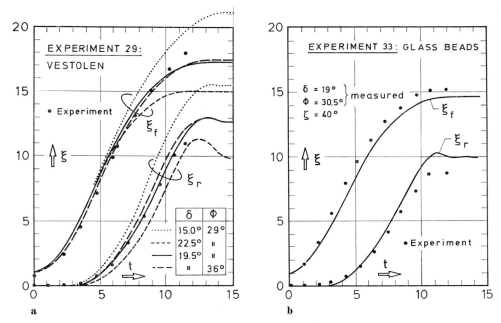

Fig. 2. Comparison of computational results as obtained by using finite difference approximations of equations (2.4) with two of Plüss' experiments. The chute was 100 mm wide and consisted of a 1.7 m long inclined and an equally long horizontal leg connected by a curved element with radius 246 mm. The slope of the inclined portion was $\zeta = 40°$ and the basal surface and the confining back wall were made of PVC while the front wall was made of plexiglass. Shown in the graphs are the results from experiments (symbols) and theoretical predictions by using equations (2.4), (continuous lines). Plotted are the dimensionless positions of the avalanche front ξ_f and the rear position, ξ_r, against time. **a** *Experiment No. 29* was performed with 500 g Vestolen particles (see Exp. No. 66 in figure caption 3 for description). Computations for internal angles of friction, ϕ and bed friction angles δ as indicated in the inset. Results are fairly insensitive to ϕ-values but depend more strongly upon δ. **b** *Experiment No. 33* was performed with 3 000 g glass beads (diameter 3.0 mm, mass density 2 730 kg m^{-3} and bulk density, corresponding to densest packing 1 780 kg m^{-3}). The friction angles are shown as insets

was tested against one of Huber's experiments of a finite mass of gravel moving down a plane inclined chute providing encouragement for continuation.

Plüss' experiments [14], first analysed in [16], and designed to study the motion of a pile of granules from initiation to runout are in a way untypical to test the suitability of the theoretical model, because the chute consisted of two straight legs (where curvature is nil) and a segment of strong curvature (where the applicability of the equations is questionable). Nevertheless, since this segment is short, the effects of the curvature cannot become appreciable and, indeed, as shown in [4] and [17], observations and computational results match satisfactorily in a variety of experiments. The results illustrated in Fig. 10 corroborate this in two further experiments, details of which can be taken from the figure caption.

A more adequate test of the theoretical model is provided by Koch's [15] experiments of the motion of a finite mass of granular material down an inclined, exponentially curved chute with small to moderate curvature. It allows verification of both sets of model equations, (2.3) and (2.4), respectively as well as of the similarity solutions. This has been done extensively by Hutter and Koch [9]. Let us summarize the findings.

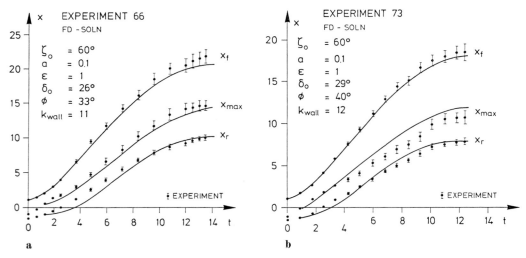

Fig. 3. Experimental measurements (symbols) and theoretical predictions (solid lines) using equations (2.3) for two granular chute experiments along an exponentially curved chute. Shown are the dimensionless positions of the avalanche front, ξ_f, the rear position, ξ_r and the position of maximum height ξ_{max} plotted against dimensionless time. The initial angle of inclination is $\zeta_0 = 60°$ and the bed geometry is described by $\zeta(\xi) = \zeta_0 e^{-0.1\xi}$. Internal angle of friction ϕ and bed friction angle δ are shown as insets. The two experimental conditions were as follows: *Experiment No. 66*: 3 litres of Vestolen particles (plastic beads, lense shaped with height 2.5 mm and diameter 4 mm, mass density 950 kg m^{-3} and bulk density, corresponding to densest packing 950 kg m^{-3}). The chute was lined with drawing paper, from [9]. *Experiment No. 73*: Same arrangement as experiment no. 66, but now quartz particles (with round shape and mean diameter approximately 3 mm, mass density 2600 kg m^{-3} and bulk density, corresponding to densest packing 1560 kg m^{-3}), from [9]

1) The theoretical model (with constant δ and ϕ) reproduces the evolution of the granular avalanche as it evolves from initiation to runout with good to satisfactory agreement, provided δ is smaller than ϕ (by approximately 3°—4°)[2]. Figure 3 provides proof of this statement; details can be taken from the figure legend.

2) When the masses of the granular piles are excessively small such that the granules can bounce extensively within the pile then the theory seems to deviate from the experiments. In this case a large amount of energy must be stored in the fluctuating motion of the particles.

3) The similarity solutions in general do not yield positions of the leading and the trailing edges in sufficient agreement with experimental observation. The deviations are particularly large when the initial velocity of the rear end of the avalanche is uphill, a situation that invalidates the applicability of the similarity solution, [11]. In these cases considerable amendments were achieved by starting computations from a time when all experimental velocities within the pile were downhill. In any event, it seems appropriate to use the similarity model only for *diagnostic* purposes and to operate for *prognostic* means with the finite-difference approximation of the original equations.

---

[2] In the experiments wall friction effects (which are absent in reality) gave rise to 'effective ϕ' so that, locally, $\phi < \delta$ could arise leading to complex valued $k_{actpass}$ (see (2.2)) and invalidating the model

2.5 Prospects

The above described state of knowledge is still incomplete and requires extension in at least the following respects.

1) It is known among field snow mechanicians that the simple Voellmy-Salm models are poor predictors for estimating runout distances when the avalanche track contains one or several (mild) bumps, so that the avalanche goes through several accelerating and decelerating phases. The moving compact mass may even split into two or more piles each of which follows its own motion and may form its own runout zone. The model equations ought to be tested against such experimental situations. This is under way.

2) Adding a viscous component to the basal drag has led to some qualitative changes in the behaviour of similarity solutions. It was thus tried (Lang [19], work in progress) to also add such a viscous drag component to the finite-difference scheme of Savage and Hutter [4], however, by its addition this scheme became numerically unstable and no complete solution could so far be found. An alternative stable scheme ought to be found, because it is known from field observations (Dr. H.-U. Gubler, personal communication) that the often observed rapid deceleration in the runout zone must be due to basal drag other than Mohr-Coulomb type.

3 Three-dimensional unconfined spreading

Whereas chute flow configurations may be adequate in many realistic situations because avalanches often move through confined gorges, these situations cannot give information about sidewise spreading when the moving mass suddenly enters an open area. The motion of a granular pile along an inclined plane (Fig. 4a) or a surface, curved in the downhill direction (Fig. 4b) are two configurations that allow theoretical and experimental study of maximum sidewise spreading. All work related to such studies is presently in progress, [19], [20], [21], [22].

3.1 Flow down inclined rough planes

The analysis performed for the plane flow can be extended to the three-dimensional situation. What results are field equations for the depth averaged velocity components parallel to the bed $u(x, y, t)$, $v(x, y, t)$ and the distribution of height $h(x, y, t)$. In the process of non-dimensionalization three ordering parameters arise

$$\varepsilon_x = H/L_x, \qquad \varepsilon_x = H/L_y, \qquad \varepsilon_{xy} = L_y/L_x,$$

and these will be chosen to have the orders of magnitude $\varepsilon_x \ll 1$, $\varepsilon_y \ll 1$ and $\varepsilon_{xy} = O(1)$. When retaining terms which are linear in ε_x and ε_y the depth averaged equations assume the form [20]

$$\frac{\partial h}{\partial t} + \frac{\partial(hu)}{\partial x} + \frac{\partial(hv)}{\partial y} = 0,$$

$$\frac{\partial(hu)}{\partial t} + \frac{\partial(hu^2)}{\partial x} + \frac{\partial(huv)}{\partial y} = Ah - B\frac{\partial(h^2/2)}{\partial x} - C \operatorname{sgn}(u) h,$$

(3.1)

$$\frac{\partial(hv)}{\partial t} + \frac{\partial(huv)}{\partial x} + \frac{\partial(hv^2)}{\partial y} = -D\,\frac{\partial(h^2/2)}{\partial y} - Ch\,\frac{v}{|u|}, \tag{3.1}$$

$$A = \sin\zeta,$$
$$B = \varepsilon_x \cos\zeta\, k^x_{\mathrm{actpass}},$$
$$C = \cos\zeta\,\tan\delta,$$
$$D = \varepsilon_y \cos\zeta\, k^y_{\mathrm{actpass}}$$

with

$$k^x_{\mathrm{act}} = \frac{2}{\cos^2\phi}\left\{1 - \sqrt{1 - (1 + \tan^2\delta)\cos^2\phi}\right\} - 1, \qquad \text{for} \qquad \frac{\partial u}{\partial x} > 0,$$

$$k^x_{\mathrm{pass}} = \frac{2}{\cos^2\phi}\left\{1 + \sqrt{1 - (1 + \tan^2\delta)\cos^2\phi}\right\} - 1, \qquad \text{for} \qquad \frac{\partial u}{\partial x} < 0,$$

$$\tag{3.2}$$

$$k^y_{\mathrm{act}} = \frac{1}{2}\left\{k^x_{\mathrm{act}} + 1 - \sqrt{(k^x_{\mathrm{act}} - 1)^2 + 4\tan^2\delta}\right\}, \qquad \text{for} \qquad \frac{\partial v}{\partial y} > 0,$$

$$k^y_{\mathrm{pass}} = \frac{1}{2}\left\{k^x_{\mathrm{pass}} + 1 + \sqrt{(k^x_{\mathrm{pass}} - 1)^2 + 4\tan^2\delta}\right\}, \qquad \text{for} \qquad \frac{\partial v}{\partial y} < 0.$$

Equation $(3.1)_1$ is the integrated mass balance equation, whereas $(3.1)_{2,3}$ are the streamwise and sidewise momentum balances with the force terms on their right hand sides. In $(3.1)_2$, Ah is the driving force — the downhill component of the gravity force. On the right, the second term in $(3.1)_2$ and the first term in $(3.1)_3$ are due to the Mohr-Coulomb type bed resistance. They are responsible for longitudinal and transverse dispersion. Finally, the last terms on the right in (3.1) are viscous drag contributions. The equations further assume that longitudinal motion is fast in comparison to sidewise motion, conditions that are violated at the initial phase of the motion.

Eqs. (3.1) are a set of partial differential equations for the unknowns $u(x, y, t)$, $v(x, y, t)$ and $h(x, y, t)$ which can be integrated, in principle, once an initial profile shape and velocity field is prescribed. A numerical integration similar to the one employed for the plane flow situation [4] has not been attempted so far and will certainly not be trivial. Instead, these equations have further be simplified by performing a second transverse averaging by guessing the transverse distributions of h, u, v as follows:

$$h(x, y, t) = h_0(x, t)\,(1 - (y/s)^2),$$

$$v(x, y, t) = v_0(x, t)\,\frac{y}{s}\,\frac{ds(x, t)}{dt}\,\frac{y}{s}. \tag{3.3}$$

$$u(x, y, t) = \hat{u}(x, t).$$

in which $s(x, t)$ is defined as the semi-span of the moving pile of mass at x. New field equations for h_0, v_0 and s are being deduced by averaging, but we have preferred to write these equations in terms of the mean fields

$$\hat{h}(x, t) = \frac{1}{s(x, t)}\int_0^{s(x,t)} h(x, y, t)\,dy, \tag{3.4}$$

$$\hat{u}(x, t) = \frac{1}{s(x, t)\,\hat{h}(x, t)}\int_0^{s(x,t)} u(x, y, t)\,h(x, y, t)\,dy \tag{3.5}$$

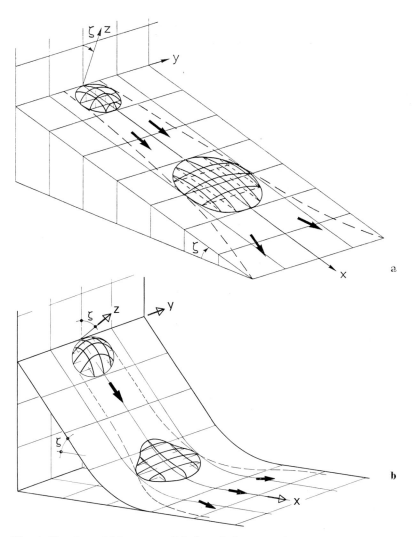

Fig. 4. Sketches of laboratory slide boards for unconfined sidewise spreading. **a** Inclined plane (ETH Zürich) **b** Curved bed (TH Darmstadt)

and then deduced the equations

$$\frac{\partial(\hat{h}s)}{\partial t} + \frac{\partial(s\hat{u}\hat{h})}{\partial x} = 0\,,$$

$$\frac{\partial \hat{u}}{\partial t} + \hat{u}\,\frac{\partial \hat{u}}{\partial x} = A - \frac{3B}{5s\hat{h}}\,\frac{\partial(s\hat{h}^2)}{\partial x} - C\,\mathrm{sgn}\,(\hat{u})\,, \qquad (3.6)$$

$$\frac{\partial v_0}{\partial t} + \hat{u}\,\frac{\partial v_0}{\partial t} = 3D\,\frac{\hat{h}(x,\,t)}{s(x,\,t)} - \frac{Cv_0}{|\hat{u}|}\,.$$

Here, $(3.6)_1$ is the mass balance, $(3.6)_2$ the streamwise and $(3.6)_3$ the sidewise momentum equation. These equations are spatially one dimensional and thus of the complexity of those in [4]. Similar integration techniques can be used to solve them.

Similarity solutions have also been constructed, [20], [21]. The system (3.1), however, does not permit them to exist exactly. However, when the viscous drag (the term in (3.3) involving C) is dropped, then they do, or else a mild approximation can be made [20]. As one might surmise from the similarity solution of the plane flow problem such similarity solutions have a pile geometry in which lines of equal hight are ellipses and the height distribution is parabolic:

$$h(x, y, t) = \frac{1}{g(t)\, f(t)} \left\{ 1 - \frac{\left(x - x_c(t)\right)^2}{g^2(t)} - \frac{y^2}{f^2(t)} \right\}. \tag{3.7}$$

$x_C(t)$ is the position of the centre of mass of the parabolic pile and $g(t)$ and $f(t)$ are the semi-axes of the ellipse, all as functions of time. For these quantities ordinary differential equations in time have been developed that describe their evolution in time. Qualitative features are discussed in [20], [21], the most important being the following:

> Along a straight inclined plane there are no rigid body solutions, in other words: There are no solutions for which $g(t) = \text{const}$ *and* $f(t) = \text{const}$, but for aspect ratios that are small ($\varepsilon_C \cong 10^{-2}$) such conditions nearly prevail.

Experiments were performed at the Laboratory of Hydraulics, Hydrology and Glaciology, ETH Zürich, involving the dispersion of a finite mass of granular material moving down an inclined rough, plane surface, having initially the shape of a circular parabolic cap. By suddenly lifting the cap (panel A of Fig. 5) the mass was released and then left to evolve. Figure 5 shows a sequence of photographs of one experiment (out of many). The panels A through H show equidistant ($\Delta t \simeq 0.04$ sec) snapshots of the granular mass of plastic beads moving down a rough plane with an inclination angle of $60°$, photographs being taken perpendicular to the inclined plane. The photographs prove that the piles deform in the course of motion from initially being circular, and then rapidly assuming the shape of 'droplets'. Thus the similarity solution is not borne out experimentally and can have only qualitative value.

In work still in progress, Savage (personal communication) has compared results from the numerical integration of eqs. (3.6) with a limited number of experiments such as those shown in Fig. 5 and reports to have obtained good agreement. Details must be deferred until the publication of his memoir.

3.2 Flow along curved beds

Report on flow along surfaces that are curved in the downhill direction is even more sketchy than that along inclined planes. Experiments performed on slide boards having the geometry sketched in Fig. 4b have been performed by Lang [19] in her doctoral dissertation, preliminary results of which are given in [22]. Lang derives the governing depth averaged equations in the curvilinear coordinasts x, y, z shown in Fig. 4b. As with the plane case of the last section, she does not integrate the emerging equations, but introduces a second, sidewise, averaging procedure to come up with equations analogous to (3.6), valid along a bed that is curved in the x-direction. She integrates these equations in space and time (x and t are the independent variables) for the geometry of the margin curve as it evolves in time and compares these with the experimental data. The limited amount of data that were analysed indicate that Mohr-Coulomb basal friction now does not suffice to match theory and observation; for that, an additional viscous drag must be added. For more details we must await the appearance of Lang's dissertation.

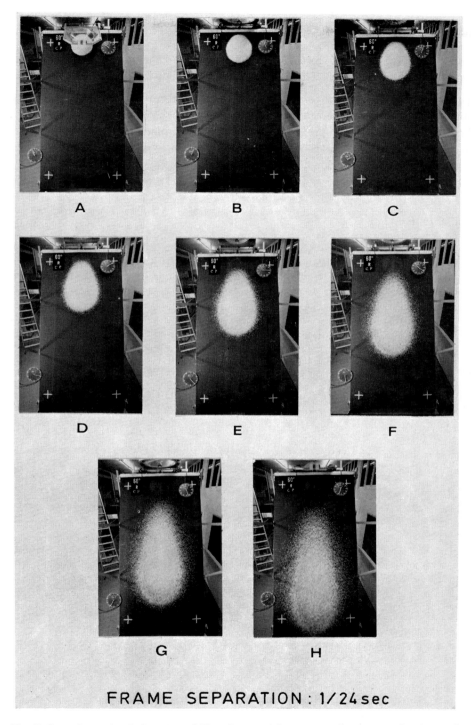

FRAME SEPARATION: 1/24 sec

Fig. 5. Snapshots of a finite mass of Vestolen particles moving down an inclined, rough plane, with $\zeta_0 = 60°$ and having initially the shape of a spherical cap, [8], [10]

4 Concluding remarks

We have reported on the status of the theoretical description of the motion of a finite mass of a cohesionless granular material. The results permit the following inferences:

— The theory predicts chute flow configurations satisfactorily if the compact mass does not split into two or more seperate entities. The study of the latter case is under way.
— Results on three dimensional spreading along an inclined plane or a curved bed indicate promising behaviour of the theory and suggest further exploration of the comparison between theory and experiment.

Acknowledgements

This work is part of a collaborative effort with Prof. S. B. Savage, McGill University. I thank Prof. Dr. D. Vischer, head, Laboratory of Hydraulics and Glaciology, ETH Zürich, for his permission to publish Fig. 5. I am grateful to Mrs. Danner for typing the text. Professors P. Haff and B. White were reviewing an earlier version of the paper; their comments are appreciated.

References

[1] Voellmy, A.: Über die Zerstörungskraft von Lawinen. Schweizerische Bauzeitung **73**, 159—162, 212—217, 246—249, 280—285 (1955).

[2] Salm, B.,: On nonuniform steady flow of avalanching snow. IUGG/IAHS General Assembly, Bern, Switzerland, IAHS Publ. No. **79**, 19—29 (1968).

[3] Perla, I. P., Cheng, T. T., McClung, D. M.: A two parameter model of snow avalanche motion. J. Glaciol. **26** (94), 197—207 (1980).

[4] Savage, S. B., Hutter, K.: The dynamics of avalanches of granular materials from initiation to runout, Part I. Analysis. Acta Mech. **86**, 201—223 (1990).

[5] Savage, S. B.: Flow of granular materials. In: Theoretical and applied mechanics (P. Germain, M. Piau, D. Caillerie, eds.), Elsevier Scientific Publ. V. V. North Holland, IUTAM 1989, 241—266

[6] Hutter, K., Szidarovsky, F., Yakowitz, S.: Plane steady shear flow of a cohesionless granular material down an inclined plane: a model for flow avalanches, Part I. Theory. Acta Mech. **63**, 87—112 (1986).

[7] Hutter, K., Szidarovsky, F., Yakowitz, S.: Plane steady shear flow of a cohesionless granular material down an inclined plane: a model for flow avalanches, Part II. Numerical results. Acta Mech. **65**, 239—261 (1986).

[8] Savage, S. B., Hutter, K.: The motion of a finite mass of granular material down a rough incline. J. Fluid Mech. **199**, 177—215 (1989).

[9] Hutter, K., Koch, T.: Motion of a granular avalanche in an exponentially curved chute: experiments and theoretical predictions. Phil. Trans. Roy. Soc. London **A 334**, 93—138 (1991).

[10] Hutter, K., Nohguchi, Y.: Similarity solutions for a Voellmy model of snow avalanches with finite mass. Acta Mech. **82**, 99—127 (1990).

[11] Savage, S. B., Nohguchi, Y.: Similarity solutions for avalanches of granular materials down curved beds. Acta Mech. **75**, 153—174 (1988).

[12] Nohguchi, Y., Hutter, K., Savage, S. B.: Similarity solutions for a finite mass granular avalanche with variable friction. Continuum Mechanics and Thermodynamics **1**, 239—265 (1989).

[13] Huber, A.: Schwallwellen in Seen als Folge von Felsstürzen. Mitteilung No. 47 der Versuchsanstalt für Wasserbau, Hydrologie und Glaziologie an der ETH, 1—122 (1980).

[14] Plüss, Ch.: Experiments on granular avalanches. Diplomarbeit, Abt. X, Eidg. Technische Hochschule, Zürich, pp. 1—113 (1987).

[15] Koch, T.: Bewegung einer Granulatlawine entlang einer gekrümmten Bahn. Diplomarbeit, Technische Hochschule Darmstadt, 1—122 (1989).

[16] Hutter, K., Plüss, Ch., Maeno, N.: Some implications deduced from laboratory experiments on granular avalanches. Mitteilung No. 94 der Versuchsanstalt für Wasserbau, Hydrologie und Glaziologie an der ETH, 323—344 (1988).

[17] Hutter, K., Plüss, Ch., Savage, S. B.: Dynamics of avalanches of granular materials from initiation to runout, Part II. Laboratory experiments (in preparation).

[18] Alean, J.: Untersuchungen über Entstehungsbedingungen und Reichweiten von Eislawinen. Mitteilung Nr. 74 der Versuchsanstalt für Wasserbau, Hydrologie und Glaziologie an der ETH, 1—217 (1984).

[19] Lang, R. M.: An experimental and analytical study on gravity driven free surface flows of cohesionless granular media. Dr. rer. nat. dissertation, Technische Hochschule Darmstadt (forthcoming).

[20] Hutter, K., Siegel, M., Savage, S. B., Nohguchi, Y.: Twodimensional spreading of a granular avalanche down an inclined plane. Part I. Theory. Acta Mech. (submitted).

[21] Hutter, K., Siegel, M.: Two-dimensional similarity solutions for finite mass granular avalanches with Coulomb and viscous-type frictional resistance. J. Glaciol. (submitted).

[22] Lang, R. M., Leo, B. R., Hutter, K.: Flow characteristics of an unconfined, non-cohesive, granular medium down an inclined curved surface, preliminary experimental results. Ann. Glaciol. 13, 146—153 (1989).

Author's address: Prof. K. Hutter, Ph. D., Institut für Mechanik, Technische Hochschule, Hochschulstrasse 1, D-W-6100 Darmstadt, Federal Republic of Germany

Aeolian Grain Transport

Volume 1: Mechanics

Edited by O. E. Barndorff-Nielsen and B. B. Willetts

(Acta Mechanica/Supplementum 1)

1991. 83 figures. Approx. 250 pages.
Soft cover DM 220,–, öS 1540,–
Reduced price for subscribers to "Acta Mechanica":
Soft cover DM 198,–, öS 1386,–
ISBN 3-211-82269-0

Prices are subject to change without notice

R. S. Anderson, M. Sørensen and B. B. Willetts:
A review of recent progress in our understanding of aeolian sediment transport

Saltation layer modelling:

R. S. Anderson and P. K. Haff: Wind modification and bed response during saltation of sand in air
I. K. McEwan and B. B. Willetts: Numerical model of the saltation cloud
M. Sørensen: An analytic model of wind-blown sand transport
M. R. Raupach: Saltation layers, vegetation canopies and roughness lengths

Observations of grain mobility:

G. R. Butterfield: Grain transport rates in steady and unsteady turbulent airflows
B. B. Willetts, I. K. McEwan and M. A. Rice: Initiation of motion of quarts sand grains
K. R. Rasmussen and H. E. Mikkelsen: Wind tunnel observations of aeolian transport rates
B. R. White and H. Mounla: An experimental study of Froude number of wind-tunnel saltation
M. A. Rice: Grain shape effects on aeolian sediment transport

Granular avalanche flow:

K. Hutter: Two- and three dimensional evolution of granular avalanche flow – Theory and experiments revisited

Springer-Verlag Wien New York